Reviews of Physiology, Biochemistry and Pharmacology 135

Springer-Verlag Berlin Heidelberg GmbH

Reviews of

135 Physiology Biochemistry and Pharmacology

Special Issue on Cyclic GMP

Editor of this Issue: G. Schultz, Berlin

With 37 Figures and 7 Tables

Springer

ISSN 0303-4240
ISBN 978-3-662-31204-9 ISBN 978-3-540-68704-7 (eBook)
DOI 10.1007/978-3-540-68704-7
Library of Congress-Catalog-Card Number 74-3674

Originally published by Springer-Verlag Berlin Heidelberg New York in 1999.
Softcover reprint of the hardcover 1st edition 1999

Production: PRO EDIT GmbH, D-69126 Heidelberg
SPIN: 10685810 27/3136-5 4 3 2 1 0 – Printed on acid-free paper

Preface

The chapters in this volume provide up-to-date information on research dealing with regulation of the biosynthesis and degradation of cyclic GMP and with the regulation of physiological systems by cyclic GMP. Research on cyclic GMP, from its inception in the 1960's to the present, has been led by a small number of laboratories concentrated in the United States and Germany. The authors of chapters in this volume, all leaders in the field, have been involved in pioneering research on cyclic GMP in one or more of those few laboratories.

Cyclic GMP was synthesized in H.G.Khorana's laboratory in 1961, well before it was known to exist in biological material. Shortly after synthetic cyclic GMP became available, Ted Rall and Earl Sutherland in Cleveland showed that it had very low potency compared to cyclic AMP in activating liver glycogen phosphorylase. In 1963, Duane Price and his colleagues in New York identified two organic phosphates in rat urine. One was cyclic AMP, which already was known to exist in urine from the work of R. W. Butcher and Earl Sutherland. The other was cyclic GMP. Three years later, James Davis, Sutherland and I confirmed the report by Price and his colleagues and showed that levels of cyclic GMP and cyclic AMP in rat urine changed independently of each other in response to altered hormonal states, implying independent metabolic pathways for the two nucleotides.

During 1969 the existence of cyclic GMP in tissues was reported by Nelson Goldberg and his colleagues in Minneapolis and by Eiji Ishikawa in Sutherland's laboratory in Nashville. The year 1969 also saw the first reports of the discovery of guanylyl cyclase by Ishikawa and me in Sutherland's laboratory, by Arnold White and Gerald Aurbach in Bethesda and by Günter Schultz, Eycke Böhme and Karin Munske in Heidelberg. In 1970, J.F.Kuo and Paul Greengard in New Haven discovered cyclic GMP-dependent protein kinase in lobster muscle. Two years later, Franz Hofmann and Guido Sold in Heidelberg demonstrated the existence of the enzyme in mammalian tissues.

During the 1970's, the properties of guanylyl cyclase were more thoroughly defined by several investigators, including David Garbers and Ted Chrisman in Nashville, Eycke Böhme and Günter Schultz in Heidelberg and by Ferid Murad and his colleagues in Charlottesville. During the 1970s and 80s, Garbers and his colleagues in Nashville and Dallas discovered the family of peptide-regulated, membrane-associated guanylyl cyclases, first in spermatazoa of marine species and then in mammalian tissues. Primarily through the work of Joe Beavo and his colleagues in Seattle, the role of cyclic GMP as a substrate and regulator of multiple isozymes of cyclic nucleotide phosphodiesterase would begin to be clarified during the 1970's and 80's.

Murad and his coworkers observed in the mid-1970's that sodium azide, included in guanylyl cyclase reaction mixtures to retard GTP breakdown, caused an increase in activity of the enzyme and went on to show that this effect was the result of the formation of nitric oxide during the metabolism of azide. Soon thereafter, Böhme, Schultz and their colleagues, Murad and his colleagues, P. A. Craven and F. R. DeRubertis in Pittsburgh and Jack Diamond in Vancouver found that many compounds capable of yielding nitric oxide activate soluble guanylyl cyclase and raise cyclic GMP in tissues. These observations, eventually converging with Rupert Gerzer's discovery in Schultzs laboratory of the heme moiety of soluble guanylyl cyclase, Robert Furchgott's discovery of endothelium-derived relaxing factor and subsequent related work by Louis Ignarro, Salvador Moncada, and Eycke Böhme and their colleagues and others would lead to our current understanding of the role of nitric oxide in signal transduction and the 1998 Nobel Prize in Medicine and Physiology for Furchgott, Murad and Ignarro.

The publication of this volume thus could not have come at a more appropriate time. Over the past decade, some of the most important advances in signal transduction research have involved cyclic GMP. Moreover, the introduction in early 1998 of sildenafil (Viagra), a selective inhibitor of a cyclic GMP phosphodiesterase, for the treatment of erectile dysfunction in men represents the first successful therapeutic application of an agent designed to alter the activity of a molecular target in a cyclic GMP metabolic pathway. The recent finding by Doris Koesling and Andreas Friebe that YC-1 potentiates the stimulatory effects of nitric oxide and carbon monoxide on cytosolic guanylyl cyclase has stimulated the development of new vasodilators which will have applications different from the commonly used nitric oxide-releasing organic nitrates. There is vigorous research on sildenafil- and YC-1-related drugs in several pharmaceutical companies and rapid growth of information about the involvement of cyclic GMP in the regulation of numerous biological processes. These factors make it likely that there soon will be more new therapeutic agents whose molecular targets will be proteins involved in the metabolism or action of cyclic GMP. Candidates for such molecular targets can be found in each chapter of this volume.

November 1998 Joel G. Hardman

Contents

Indexed in Current Contents

Mechanisms of Regulation and Functions of Guanylyl Cyclases

D. C. Foster, B. J. Wedel, S. W. Robinson and D. L. Garbers

Howard Hughes Medical Institute and Department of Pharmacology,
University of Texas Southwestern Medical Center, Dallas, TX 75235-9050

Contents

Abbreviations

NO, nitric oxide; CO, carbon monoxide; ANP, atrial natriuretic peptide; AMPPNP, 5'-adenylylimidodiphosphate; ATPγS, adenosine 5'-O-thiotriphosphate; 2'd3'AMP, 2'-deoxyadenosine 3'-monophosphate; sGC, soluble guanylyl cyclase; mGC, membrane guanylyl cyclase; AC, adenylyl cyclase; STa, heat-stable enterotoxin of E. coli; PKC, protein kinase C; GC-A-G, guanylyl cyclase A-G; TM, transmembrane domain; ECD, extracellular domain; KHD, kinase homology domain; DD, dimerization domain; CHD cyclase homology domain; HBD, heme binding domain; ODQ, 1*H*-[1,2,4]oxadiazolo[4,3-*a*]quinoxalin-1-one.

1 Introduction

Guanylyl cyclases exist in both the soluble and membrane fractions of most tissue homogenates, probably across all animals. The conservation of amino acids within "signature" domains is pronounced, and thus a cyclase homology domain (CHD), known to possess either adenylyl or guanylyl cyclase activity, and an amphipathic region found in both soluble and membrane forms of guanylyl cyclase are easily recognized at the primary amino acid level. In addition, amino terminal regions of the soluble forms, suggested to bind heme, and a protein kinase homology domain (KHD) of the membrane forms, retain a high degree of conservation at the amino acid sequence level. The most variable region of the cyclases is within the putative extracellular, ligand binding domain of the membrane forms. However, even in this case certain cysteine residues are highly conserved.

A recent proliferation in the discovery of new cyclase forms has significantly expanded the database, allowing a rather exhaustive determination of those amino acids that are invariant or highly conserved in all cyclases. However, most guanylyl cyclases remain orphan receptors, and the identification of the putative ligands remains central to our understanding the functions of each enzyme.

2 General Features of Guanylyl Cyclases

Cyclase Homology Domain. All animal adenylyl and guanylyl cyclases contain a region (cyclase homology domain, CHD) that is highly conserved (Fig. 1). Based on various studies of fragments of adenylyl cyclases, membrane

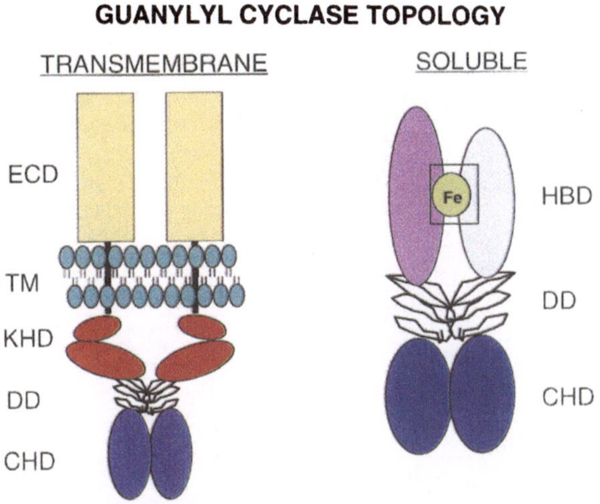

Fig. 1. Signature Domains of Guanylyl Cyclases. A. Membrane guanylyl cyclases are comprised of an extracellular domain (ECD), a single putative transmembrane segment (TM), and an intracellular protein-kinase homology domain (KHD), dimerization domain (DD) and cyclase homology domain (CHD). B. Soluble guanylyl cyclases are comprised of N-terminal domains suggested to bind a prosthetic heme moiety (illustrated is the central Fe^{2+}; heme binding domain, (HBD)), a dimerization domain and a carboxyl terminal guanylyl cyclase catalytic domain

guanylyl cyclases, and soluble guanylyl cyclase heterodimers, the CHD is known to represent the catalytic heart of the protein (Harteneck et al. 1990, Nakane et al. 1990, Thorpe and Morkin 1990, Thompson and Garbers 1994, Garbers et al. 1994, Tang and Gilman 1995). Activity has been undetectable for monomeric species of any of the cyclases, and thus the minimal catalytic unit appears to be that of a dimer for either adenylyl or guanylyl cyclases. For adenylyl and soluble forms of guanylyl cyclase the minimal active unit appears to be that of a heterodimer, while the membrane forms of guanylyl cyclase studied form homomers (Garbers et al. 1994). The membrane forms of guanylyl cyclase and the β-subunit of the soluble guanylyl cyclases appear to more closely resemble the C_2 region of adenylyl cyclases, while the CHD primary sequence of adenylyl cyclase C_1 domain appears to more closely resemble that of the α-subunit of soluble guanylyl cyclases (Fig. 2).

The solving of the crystal structures for the C_2/C_2 homomer (Zhang et al. 1997) and C_1/C_2 heteromer (Tesmer et al. 1997) have proven invaluable for modeling the active site of the guanylyl cyclases and subsequently predicting those amino acids that dictate nucleotide specificity (Figs. 2 and 3). To determine whether the substrate specificity of adenylyl and guanylyl cyclases

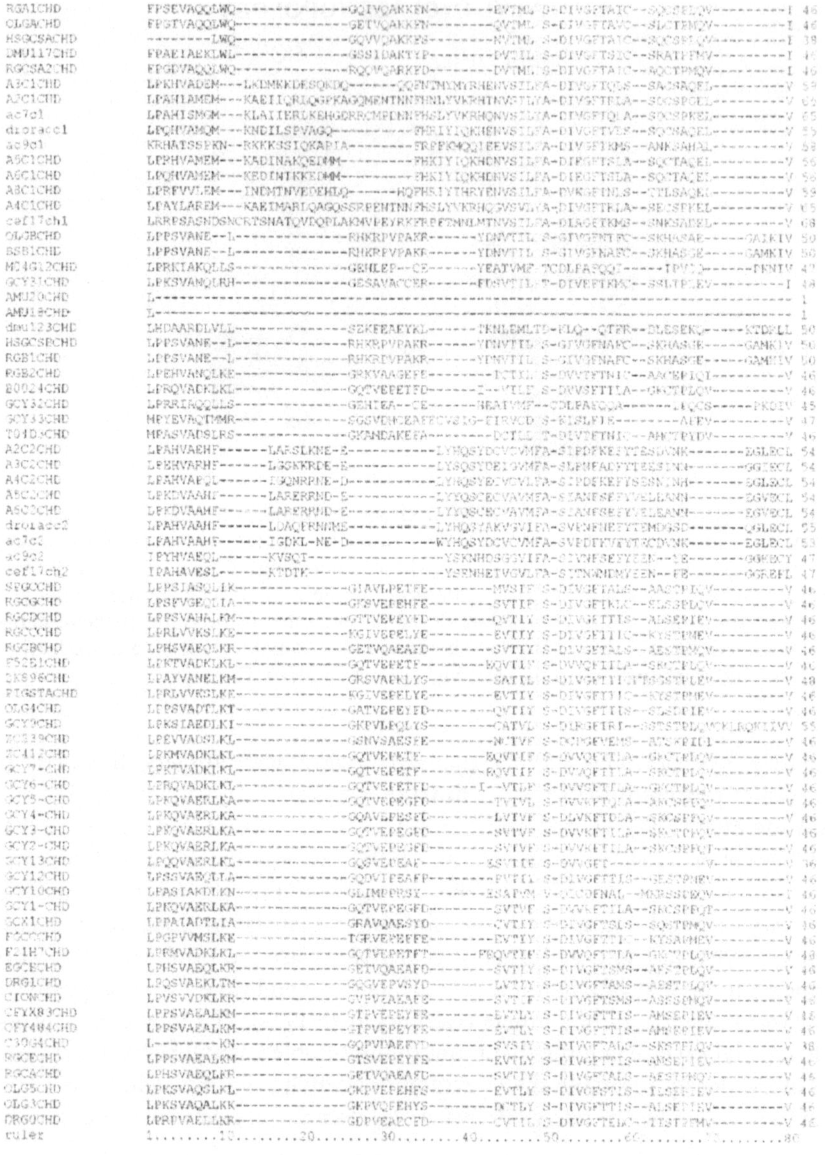

Fig. 2.

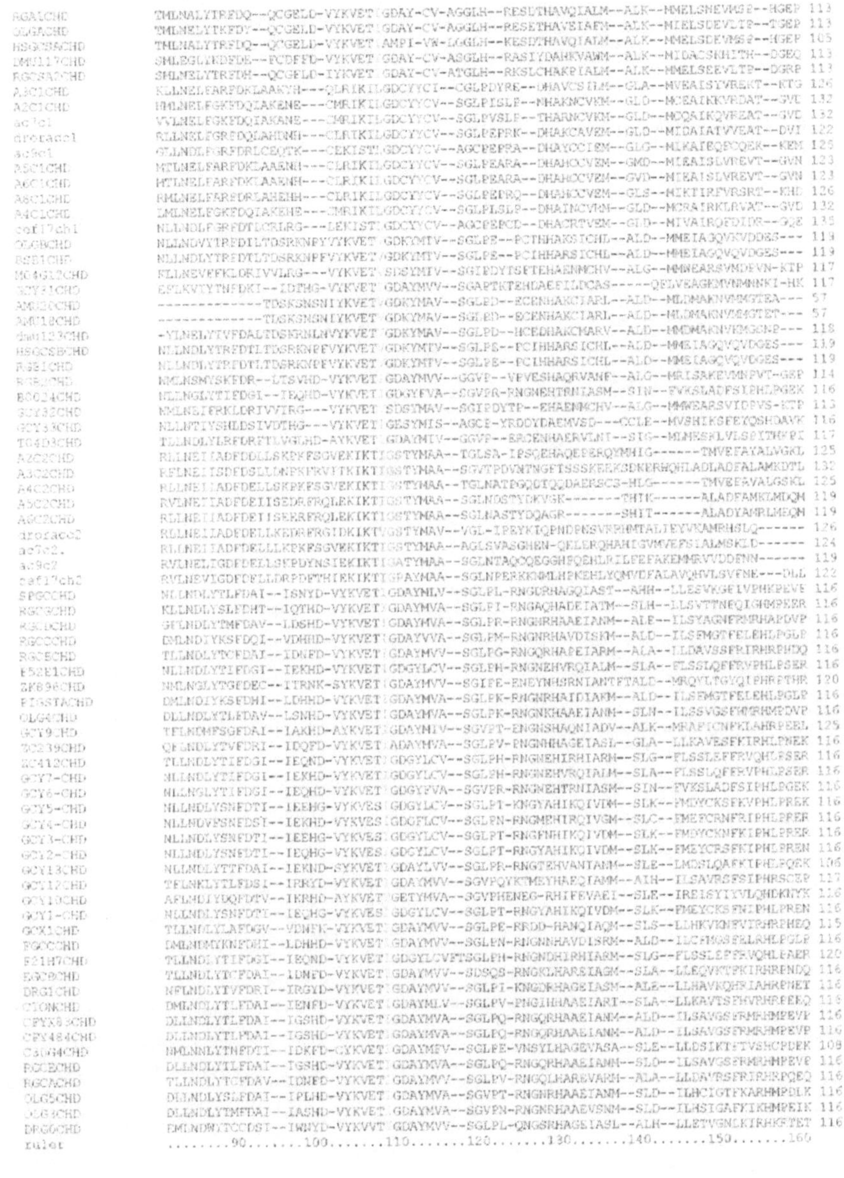

Fig. 2.

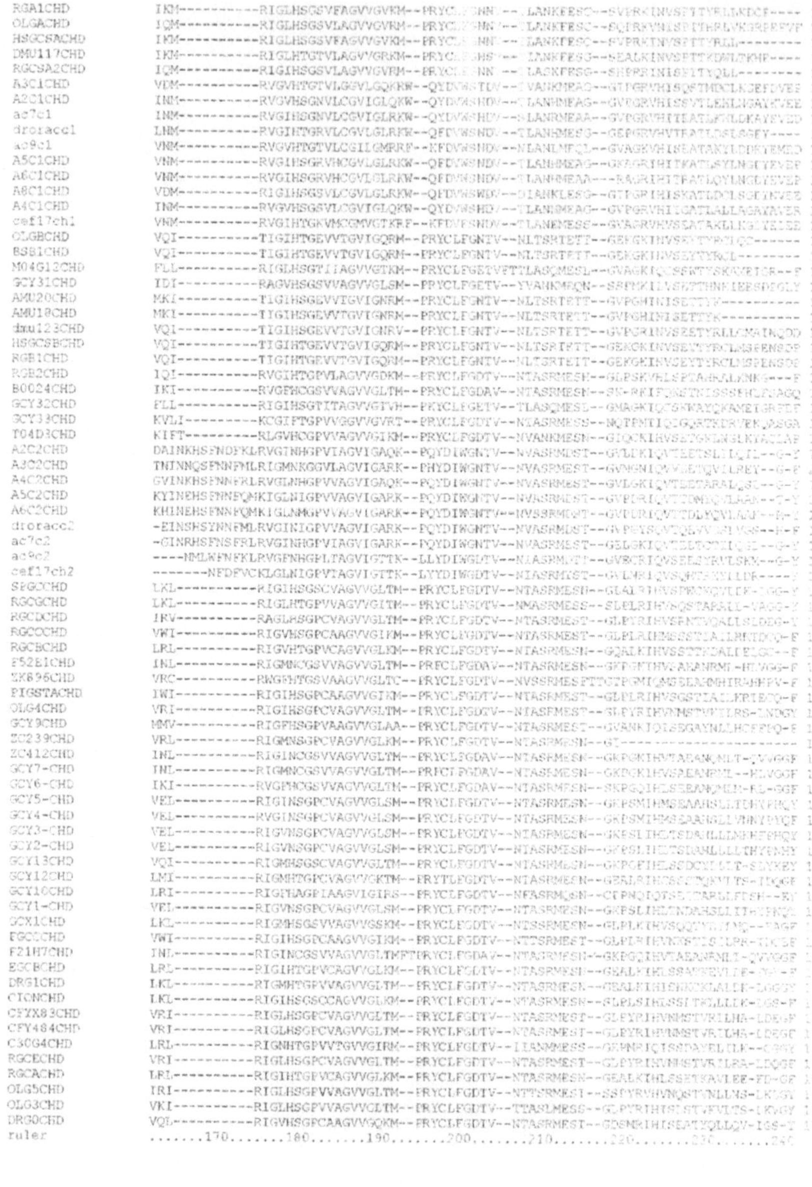

Fig. 2. Sequence Alignment of Cyclase Domains. Sequences encoding confirmed guanylyl or adenylyl cyclases or sequences identified through database searching and predicted to encode a purine cyclase were aligned. Amino acids that contribute to nucleotide substrate specificity or are conserved with amino acids that contribute to substrate specificity are colored dark blue and red, indicative of specificity for

```
RGA1CHD     -------------------------------------------------------------------------------- 173
OLGACHD     IPR----------------------------------------------------------------------------- 180
HSGCSACHD   -------------------------------------------------------------------------------- 161
DMU117CHD   -------------------------------------------------------------------------------- 173
RGC8A2CHD   -------------------------------------------------------------------------------- 169
A3C1CHD     GDG--------------GSRC-----DYL-DEKGIE--------TYLIIASKPEVK--KTAQNGLN-------------- 226
A2C1CHD     GDG--------------EIRD-----PYL-KQHLVK--------TYFVINPKGERRSP--QH--LFRPRHTLDGAKMRAS 244
ac7c1       GHG--------------QQRD-----PYL-KEMNIR--------TYLVIDPRSQQPPPPSQH--LPRPKGD-AALKMRAS 245
droracc1    -------------------------------------------------------------------------------- 182
ac9c1       GRVIE---------RL-GQSVVA---DQL-KGLK----------TYLISGQRAKESHCSCAE-ALLSGFEVIDDSRESSG 244
A5C1CHD     GCG--------------GERN-----AYL-KEHSIE--------TFLILRCTQ--K--RKEEKAMI-------------- 221
A6C1CHD     GRG--------------GERN-----AYL-KEQHIE--------TFLILGASQ--K--RKEEKA---------------- 218
A8C1CHD     GHG--------------KERN-----EFL-RKHNIE--------TYL--------------------------------- 209
A4C1CHD     ADM--------------EHRD-----PYL-RELGEP--------TYLVIDPWAEEEDEKGTERGLLSSLEGHT---MRPS 245
cef17ch1    GPDYDGPL-----RM-QVQGTE----PRV-KPESMK--------TFFIK------------------------------- 229
OLGBCHD     -------------------------------------------------------------------------------- 177
BSB1CHD     -------------------------------------------------------------------------------- 175
M04G12CHD   EFSPRGR-------------------INV-KGRGDVE-------TYFLMR------------------------------ 204
GCY31CHD    QFERREE-------------------IEI-KDDQTIQ-------TFFVVSRHGP-------------------------- 208
AMU20CHD    -------------------------------------------------------------------------------- 111
AMU18CHD    -------------------------------------------------------------------------------- 111
dmu123CHD   SFHLEYRGP-----------------VIM-KGKPTPMDC-----WFLTRA------------------------------ 209
HSGCSBCHD   QFHLEHRGP-----------------VSM-KGKKEPMQV-----WFL--------------------------------- 207
RGB1CHD     QFHLEHRGP-----------------VSM-KGKKEPMQV-----WFLSRK------------------------------ 210
RGB2CHD     EIVRRGE-------------------IEV-KGKGKM-------TTYFLIQNL---------------------------- 200
B0024CHD    IHLSEEANQMLM-RL-GGFTTEPRGEVII-KGKGVM-------ATYWLL------------------------------- 218
GCY32CHD    --SPRGR-------------------IDV-KQRGLTE-------TYFLTRSL---------------------------- 200
GCY33CHD    FRIKPKGN------------------VFV-KGKGDMR-------VYEIE------------------------------- 204
T04D3CHD    TSAIGNYRSYRTPNNFFTKLLLCKHQTKIQKTGGRQ------AGTYACL------------------------------- 225
A2C2CHD     TCTCRGI-------------------INV-KGKGDLK-------TY---------------------------------- 215
A3C2CHD     RFVRRGP-------------------IFV-KGKGELL-------TF---------------------------------- 222
A4C2CHD     TCYSRGV-------------------IKV-KGKGQLC-------TY---------------------------------- 215
A5C2CHD     QLECRGV-------------------VKV-KGKGEMM-------TY---------------------------------- 209
A6C2CHD     QLECRGV-------------------VKV-KGKGEMT-------TY---------------------------------- 209
droracc2    EFRCRGT-------------------IKV-KGKGDMV-------TYFL-------------------------------- 217
ac7c2       SCECRGL-------------------INV-KGKGELR-------TYFV-------------------------------- 215
ac9c2       DFDYRGT-------------------VNV-KGKGQMK-------TYLY-------------------------------- 207
cef17ch2    EFEFRDH-------------------IEV-KGIDGGMD------TYL--------------------------------- 206
SPGCCHD     ELEDRGL-------------------VPM-NGKGEIH-------TFWLL------------------------------- 200
RGCGCHD     HLQKRGT-------------------ISV-KGKGEQT-------TFWLTGK----------------------------- 202
RGCDCHD     KIDVRGQ-------------------TEL-KGKGLEE-------TYWLTGK----------------------------- 203
RGCCCHD     LYEVRGE-------------------TYL-KGRGTET-------TYWLTGMK---------------------------- 204
RGCBCHD     QLELRGD-------------------VEM-KGKGKMR-------TYWLLGER---------------------------- 203
F52E1CHD    DTESRGE-------------------VII-KGKGVME-------TFWLTGQG---------------------------- 203
ZK896CHD    TTTERGE-------------------VQV-KGKGTCR-------TFWLEDRV---------------------------- 210
PIGSTACHD   LYEVRGE-------------------TYL-KGRGTET-------TYWLTGVK---------------------------- 204
OLG4CHD     KIDVRGK-------------------TEL-KGKGIEE-------TYWLVG------------------------------ 202
GCY9CHD     QMVERGK-------------------IEV-KV------------------------------------------------ 200
ZC239CHD    -------------------------------------------------------------------------------- 158
ZC412CHD    RTESRGE-------------------VII-KGKGVME-------TYWL-------------------------------- 200
GCY7-CHD    DTESRGE-------------------VII-KGKGVME-------TFWLT------------------------------- 200
GCY6-CHD    TTEPRGE-------------------VII-KGKGVMA-------TYWLLKMDE--------------------------- 204
GCY5-CHD    ETSSRGE-------------------VII-KGKGVME-------TFWVLGKTD--------------------------- 206
GCY4-CHD    ETNSRGE-------------------VII-KGKGVME-------TYWLLGKM---------------------------- 205
GCY3-CHD    ETNSRGE-------------------VII-KGKGVME-------TFWVLGRS---------------------------- 205
GCY2-CHD    DTSSRGE-------------------VII-KGKGVME-------TFWVH------------------------------- 202
GCY13CHD    NTESRGE-------------------VII-KGKGVMQ-------TYWLLGMKEESA------------------------ 198
GCY12CHD    LLEERGS-------------------LAI-KGKGQMT-------TYWLNGRA---------------------------- 205
GCY10CHD    KFVKRGI-------------------VHV-KGKGNAARLKICCETFETHSIDL--------------------------- 211
GCY1-CHD    ETSSRGE-------------------VII-KGKGVME-------TFWVHG------------------------------ 203
GCX1CHD     KLELRGS-------------------VEM-KGKGMQT-------TYWLRGYKDVE------------------------- 205
FGCCCHD     QYEVRGE-------------------TYL-KGKGPEI-------TYWLT------------------------------- 201
F21H7CHD    RTESRGE-------------------VII-KGKGVME-----FTTFWLL------------------------------- 209
EGCBCHD     DLQLRGD-------------------VEM-KGKGKMR-------TYWLLG------------------------------ 201
DRG1CHD     ITEKRGL-------------------VNM-KGKGDVV-------TWWLTGAN---------------------------- 204
CIONCHD     IMQHRGE-------------------VEM-KGKGKMD-------TYFLLGEDP--------------------------- 204
CFYX83CHD   QTEVRGR-------------------TEL-KGKGAED-------TYWLV------------------------------- 201
CFY484CHD   QTEVRGR-------------------TEL-KGKGAED-------TYWLV------------------------------- 201
C30G4CHD    VTEQREK-------------------IVL-KNK--LE-----VMTYWMNDY----------------------------- 194
RGCECHD     QMECRGR-------------------TEL-KGKGVED-------TYWLVGR----------------------------- 203
RGCACHD     ELELRGD-------------------VEM-KGKGKVR-------TYWLLGER---------------------------- 203
OLG5CHD     KIDVRGL-------------------TEL-KGKGIET-------TYWLVGK----------------------------- 203
OLG3CHD     HIETR----------------------KA-QVKGTED-------TYWLMGR----------------------------- 200
DRG0CHD     VCIERGL-------------------TSI-KGKGDMR-------TYWLTKR----------------------------- 202
ruler       .......250.......260.......270.......280.......290.......300.......310.......320
```

guanosine or adenosine triphosphate, respectively. Amino acids colored green interact with forskolin. Light blue amino acids are conserved with those that interact with forskolin and may contribute to a putative regulatory pocket within the cyclase homology domain. Amino acids colored magenta interact with $G_{s\alpha}$. The identity of the sequences is provided in the Appendix

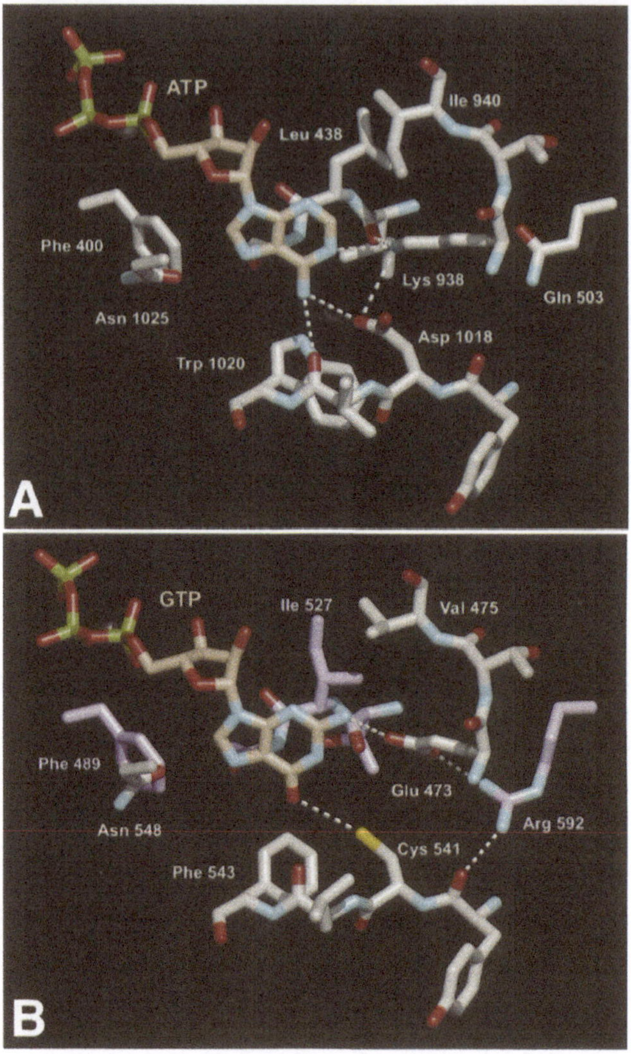
ATP
Ile 940
Leu 438
Phe 400
Lys 938
Gln 503
Asn 1025
Asp 1018
Trp 1020
A
GTP
Ile 527
Val 475
Phe 489
Glu 473
Asn 548
Arg 592
Cys 541
Phe 543
B

can be changed to the opposite nucleotide, point mutations have been made in soluble forms of guanylyl cyclase and adenylyl cyclase. The advantage of initiating such studies with these enzymes is that they form heteromers; thus single mutations can be designed within each subunit as opposed to the necessary double mutations required for the homomeric membrane forms. A change of Asp473 to Arg (found in the KVETIG consensus in guanylyl cyclases) and Cys541 to Asp (found in the MPRYCL guanylyl cyclase consensus) in the β-subunit, and Arg592 to Gln (of the MPRYCL consensus) in the α-subunit of the soluble guanylyl cyclase results in complete conversion of the enzyme to an adenylyl cyclase, while maintaining NO sensitivity (Sunahara et al, 1998). The reverse mutations in adenylyl cyclase do not result in complete conversion, although GTP is now an effective substrate and the sensitivity to G_s and forskolin is maintained. Similar studies have also identified amino acids in a retinal guanylyl cyclase (GC-E) that contribute to substrate specificity. When amino acids in GC-E that are homologous to Asp473 and Cys541 of the soluble guanylyl cyclase were mutated to the corresponding amino acids of adenylyl cyclase, ATP became a substrate, albeit with a substantial increase in the apparent Km, while GTP no longer served as a substrate for the enzyme (Tucker et al. 1998). ATP continued to serve as a substrate for the mutant transmembrane guanylyl cyclase at concentrations of the nucleotide that would have complexed a majority of Mg^{2+} present in the reaction. In fact, there was no apparent inhibition of catalysis due to a lack of free metal, a co-factor predicted as essential for guanylyl cyclase activity. It appears, therefore, that the mutations not only altered substrate specificity but abolished a free metal binding site.

The solving of the adenylyl cyclase structure predicted the binding site for forskolin, a site somewhat equivalent to that predicted for ATP binding. Only one forskolin binding site was identified in the structure, confirming the equilibrium binding data of Dessauer and Gilman (1997). Modeling of

Fig. 3. Structural model of the purine binding pockets of adenylyl and soluble guanylyl cyclase. A. Stick model of the adenine binding pocket in adenylyl cyclase. The model for ATP was based on the structure of $G_{s\alpha}$-VC_1-IIC_2 complexed with 2'deoxyadenosine-3'-monophosphate (2'd3'AMP) (Tesmer et al. 1997, Sunahara et al. 1998). Putative hydrogen bonds are shown as dotted white lines. Protein carbons are gray, nitrogens are blue, oxygens are red, phosphorus atoms are green and sulfurs are yellow. Carbons in ATP are copper. B. Stick model of the guanine binding pocket of soluble guanylyl cyclase. The coloring scheme is the same as for "A" except that carbon atoms of residues that belong to the α_1 subunit and their labels are colored mauve. Carbon atoms belonging to the β_1 subunit are gray. Figures "A" and "B" were created as described in (Sunahara et al. 1998) and are published with permission

the soluble guanylyl cyclase catalytic domain based on the adenylyl cyclase structure also predicted one active site per heteromer (Liu et al. 1997). Interestingly, the binding site for forskolin is predicted to be retained as a potential binding pocket in the guanylyl cyclase CHD. Forskolin, however, which markedly stimulates adenylyl cyclases, does not stimulate guanylyl cyclases. Thus, an important question that arises from analysis of the structural models of guanylyl cyclases is whether the site that is homologous to the forskolin binding region is an unidentified regulatory region in the guanylyl cyclases (see Wedel and Garbers, 1998).

Recently, several molecules have been described that selectively modulate the activity of soluble guanylyl cyclase. That such exogenous molecules possess the capacity to modulate soluble guanylyl cyclase activity also raises the possibility that unidentified endogenous molecules may have similar effects. By studying the mechanism of action of these molecules, insight into potential mechanisms of action of endogenous regulators of soluble guanylyl cyclase may be gained.

1*H*-[1,2,4]Oxadiazolo[4,3-*a*]quinoxalin-1-one, (ODQ), inhibits the stimulation of soluble guanylyl cyclase while possessing no detectable effect on signaling by transmembrane guanylyl cyclases (Garthwaite et al. 1995). The inhibition seems to be noncompetitive with respect to substrate. Incubation of soluble cyclase with ODQ results in a shift of the Soret band of the heme suggesting that ODQ oxidizes the Fe^{2+} of the enzyme to Fe^{3+}, which is unresponsive to NO stimulation (Schrammel et al. 1996). Although this molecule likely acts at a site distinct from the cyclase catalytic domain, it may prove useful in elucidating not only the physiological roles of soluble cyclases but also the regulatory features of this enzyme.

Another important compound is the recently described benzyl indazole derivative, YC-1. This molecule has the fascinating property of not only stimulating soluble guanylyl cyclase, but rendering the enzyme markedly sensitive to stimulation by CO (see below) (Friebe et al. 1996, Stone and Marletta, 1998). The site of YC-1 binding is not known, however, it is likely that it does not interact with the heme as incubation with YC-1 does not change the Soret absorption of soluble guanylyl cyclase either under basal or stimulated conditions (Friebe and Koesling 1998). YC-1, therefore, has been suggested to bind to an allosteric site resulting in stimulation. A decreased dissociation rate of the ligands, NO or CO, leading to enhanced stimulation by both activators has been suggested (Friebe and Koesling, 1998), although the results with CO have not been confirmed by Stone and Marletta (1998).

Structural studies of adenylyl cyclase have also provided detailed insight into the sites of interaction of numerous cyclase regulators including substrate, forskolin and the GTP-binding protein (G-protein), $G_{s\alpha}$ (Tesmer et al.

1997). The sites of interaction of $G_{s\alpha}$ with adenylyl cyclase are mapped primarily to regions of the C_2 domain of the cyclase. Although multiple contacts are made between the proteins, one of the primary interaction sites is located in a loop region between helix 3 and beta strand 4 of the C_2 domain of adenylyl cyclase. A similar conserved loop region is not found in the guanylyl cyclases, consistent with G-proteins not serving as direct regulators of guanylyl cyclase (Fig. 2).

Protein Kinase Homology Domain. The KHD is present in all membrane forms so far discovered but is absent in the soluble subunits. The function of the KHD has not been fully resolved, although it appears to act at least in part as a negative regulator of the CHD. The binding of ATP together with extracellular ligand appears to relieve this inhibition (Chinkers and Garbers 1989). Based on primary sequence alignment, the KHD closely resembles the catalytic domain of all protein kinases (Fig. 4). Most of the guanylyl cyclases, however, do not contain a glycine-loop, important in anchoring the phosphate groups of ATP, and none of the cyclases contain the Asp predicted to function as the catalytic base (Taylor et al. 1992). The HRD consensus sequence of protein kinases is replaced with HGX (where X tends to most often be Arg or Asn). Not only the guanylyl cyclases, but also the JH2 domain of the JAK family of protein kinases and a group of receptor-like molecules found in the plant species *Arabidopsis thaliana*, contain the unique HGX sequence instead of HRD (Fig. 5). No function as yet has been ascribed to this domain of the JAKs, however, that KHDs comprise the entire intracellular domain of a family of putative transmembrane receptors identified in *Arabidopsis*, strongly suggests that this highly conserved domain may itself have a signaling function.

With respect to functions of the KHD, the two natriuretic peptide receptors (GC-A and GC-B) have been the most carefully studied. In broken cell preparations, both the peptide ligand and an adenine nucleotide appear required for activation of the cyclase (Kurose et al. 1987, Chang et al. 1990, Chinkers et al. 1991). The generality of the adenine nucleotide requirement is not known, but adenine nucleotides also potentiate heat-stable enterotoxin (STa)-stimulated GC-C activity (Vaandrager et al. 1993), and photoreceptor guanylyl cyclase activity in the presence of a Ca^{2+}- dependent regulator protein (GCAP-2) (Laura et al. 1996). Evidence that the KHD represents the site of adenine nucleotide binding is indirect, but is supported by the following observations: (1) deletion of the KHD in GC-A results in an apparently constitutively active cyclase (Chinkers and Garbers 1989). (2) Various point mutations within the KHD of GC-A destroy ligand/adenine nucleo-

```
rgcakin    -Q-------------------------------LEKELVSE--LWRVRWEDLQPSSLERHLRSAGSRLTLSGRGSNYGSLLTTEGQFQVFAKTAYY---- 62
rgcbkin    LM-------------------------------LEKELASM--LWRIRWEELQFGNSDRYHKGAGSRLTLSLRGSSYGSLMTAHGKYQIFANTGHF---- 63
egcbkin    LK-------------------------------LEKELAGM--LWRIRWEELQFESPNKYHKCAGSRLTISQRGSSYGSLITAHGKYQLFAKTGYF---- 63
cionkin    MNIMIVNSAAGGEIFDHCIGVKDPFNETTTRRLLQQILEAVDFLHSNNIVHLDLKPQNILLTEGGVDGDIKLVDFGLSKILAQEIEIREILGTPDYVAPE 100
gcx1kin    INE---------------VQFRLPLND---RRVSSPSSEAT---RKKNSNEEDPENGAK------KSLSLKNRKLSFGMV--------------SFKSGS 59
spgckin    AYEAALDS----------------------------------LVWKVDWSEVQTKA----TDTNSQGFSMKNMVMSAISVISNAEKQQIFATIGTY---- 58
rgcgkin    -----------GKVQNH----------------------PGDTWWQIHYDSITLLPQ---HKPSHRGTPMSRCNVSNASTVKISADCGSFAKT------- 57
drg0kin    ----------------------------------------------------------------------------------------------------
drg1kin    I-ELEIEG----------------------------------LLWKIDPNEI--KG----YSGNEIVSSPSKVSLMSAQSYGSRWTNQFVTSTGRL---- 55
zc239kin   KRY---------RYERRLHSLFFMIDRNQIILKKHTNLMSQQSLRSM--ASIHGSVVAASQTLRDSHFFIEDYNNASSINASSIFNTGSTARAGPFGPIP 89
gcy12kin   -------------FEKEL-SMIWKIDPYEVR-------------RVV--GGVNN---ESTASLMQSDVMQ--------------FAKTKTPW-------- 46
rgcckin    R--------------------------------------RDHELRQKKWSHIPSENIFPLETNETNHVSLKIDDDR------------------------ 38
humstakin  RK-----------------------------------YRKDYELRQKKWSHIPPENIFPLETNETNHVSLKIDDDK------------------------ 41
pigstakin  RK-----------------------------------YKREYALRQKKWSHIPPENIFPLESNETNHVSLKIDDDR------------------------ 41
fgcckin    ---------------------------------------LENELRQKKWSHISAEKILPLNMTETSHVSLKIDDDK------------------------ 37
rgcdkin    LQQL--------RLLRGPHRILLTPQELT---FLQRTPSRR----RPH-----VDSGS----ESRSVVDGGSPQSVIQGSTRSVPAFLEHTNVALY---- 72
rgcekin    ---LLH--M---QMVSGPNKIILTLEDVT---FLHPQGGSS----R--------KVAQ----GSRSSLATRSTS---DIRSVPSQP-QESTNIGLY---- 65
rgcfkin    ---INK--I---QLIKGPNRILLTLEDVT---FINPHFGSK----RGSRASVSFQIIS----EVQSG---RSPRLSFSSGSLTPAT-YENSNIAIY---- 73
olg3kin    KKSANL--A----------KLILTLDDIV---FIDTQVSRK----K-------LNDES----IMRSLLEIKTPLRSIA-RSYILTT-PESSNIGIL---- 64
olg4kin    ---LQQ--I---QLVKGPNRILLTLEDLT---FINPQLSKK---------EITLDDLS----ESKSALDDKSADPSHSMNSMQTAI-HENSSVAVY---- 71
olg5kin    EGRVGF--SFGGDGNGGPSKVVLTLDDLV---LINTQVSKR----K-------LNDES----IVKSQMDLKTPHHSVSGRSYLAST-PDSSNIAVY---- 75
cfy484kin  RHRLLH--I---QMVSGPNKIILTLDDVT---FLHPHGGST----R--------KVVQ----GSRSSLAARSTS---DIRSVPSQP-LDNSNIGLF---- 68
cfyx83kin  RHRLLH--I---QMVSGPNKIILTLDDVT---FLHPHGGST----R--------KVVQ----GSRSSLAARSTS---DIRSVPSQP-LDNSNIGLF---- 68
zk896kin   RYC--------------ENKQLEKMP------------------WRIFHDDLQFIDEEQVKSMMSVGSVTTKLSNIQTGQ--KQHAIIG----VNFTTHT 62
gcy9kin    --------------------LLYDMT------------------WRIPRESIKML-EGKSKSEHSLASKSQSSGSFSGSMNSKQNGLIAAKQAVSNGVKL 61
gcy4-kin   -----------------------------------VEQSRLHSEWQIHAIKLRLPMHRRASKSS-QESETESASETENFTSKSG-----DTMTSEFKETY 59
gcy5-kin   -----------------------------------AEQERLNSEWQIPSIQLIMPQKEKR-KPN-SRRSLQSGPST--ITGESK-----MTIDGGFHENY 56
gcy1-kin   ---------------------------------KRAERARINAEWQVPFAKLIESEKQVRGKGA-SRRSLQSAPSI--STGHSG-----VTTVSDFCENY 59
gcy2-kin   -----------------------------------AEQARTNAEWQVPFVNLMESEKQIRSNAT-SRRSLQSAPSI--STGHSG-----VTTVSDFCENY 57
gcy3-kin   -----------------------------------EEQARINSEWQIPFVKLRELER--KSKGT-SKRSLQSAPST--ITGESK-----VSTGSEFCENY 55
gcy13kin   ---------------------------------RRDEELRLDDQWIVPHGMLQSI-IKGRKESHHSSRSLQSNSTTTTGTTGIS-----SRSVF-FPETE 60
f21h7kin   ------------------------------------EMERQDLLWQVAFTFVELQQVQSKSRAEASMHSFASGPST---STKMT-----VESR---TETT 53
zc412kin   IA-MAKSRKLKSQVFPFSITLLLNF------RMKRKEMERQDLLWQVP--FIELQQVQSKSKAEASMHSFASGPST---STKIT-----VESR---SETI 80
f52e1kin   ---------------------------------RRQEIERLNRLWQIPFIHLHQINSKQKGKGEHSVRSLQSGTSTLSSRTTVS-----FKT-----ESR 57
gcy7-kin   -----------------------------------QEIERLNRLWQIPFIHLHQINSKQKGKGEHSVRSLQSGTSTLSSRTTVS-----FKT-----ESR 55
b0024kin   ---------------------------------KQQEVERQNALWQIPFKSMMTVT--KKGKGEHSMRSISSVPSTISSTRSST-----LSEV---GETR 57
gcy6-kin   ----------------------------------QQEVERQNALWQIPFKSMMTVT--KKGKGEHSMRSISSVPSTISSTRSST-----LSEV---GETR 56
ruler      1.......10........20........30........40........50........60........70........80........90.......100
```

Fig. 4

```
rgcakin    -------------------------------KGNLVAVKRVN---RK--RIELTRKVLFELKHMRDVQNEHLTRFVGACTDPPNIC-----------ILT 115
rgcbkin    -------------------------------KGNVVAIKHVN---KK--RIELTRQVLFELKHMRDVQFNHLTRFIGACIDPPNIC-----------IVT 116
egcbkin    -------------------------------KGNLVAIKHVN---KK--RIELTRQVLFELKHMRDVQFNHLTRFIGACIDPPNIC-----------IVT 116
cionkin    VLNFEPISTL-TDIWYVLNKYLP----IFYTQGKLVTVKLMS---RR--KIDLTKKFLMELKHMRDVSHDHITRFEGACLDPR-IC-----------VLT 178
gcx1kin    GGSVETIAQN-NTQIYTKT---A----IF--KGVVVAIKKLNIDPKKYPRLDLSRAQLMELKKMKDLQHDHITRFTGACIDFPHYC-----------VVT 138
spgckin    -------------------------------RGTVCALHAVH-KN----HIDLTRAVRTELKIMRDMRHDNICPFIGACIDRPHIS-----------ILM 111
rgcgkin    ---------HQ-DEELF----YAP----VGLYQGNHVALCYIGEEAE--ARIK-KPTVLREVWLMCELKHENIVPFFGVCTEPPNIC-----------IVT 126
drg0kin    ----------------------------------------------------------------------------------------------------
drg1kin    -------------------------------RGAVVRIKELKFPR----KRDISREIMKEMRLLRELRHDNINSFIGASVEPTRIL-----------LVT 109
zc239kin   GFGGVTGASE-DEKWHQIPDFGVGL-----YEGRTVALKRI-Y-RSD---VEFTRSNRLEIAKLQESVNSNVIEFVGMVVQSPDV-----------FVVY 167
gcy12kin   --------------WSKAPVQGTGMRGLASYKGTLVGLKDLMYGRKP---KDLTREAKKELRAMRQLAHPNVNNFLGIIVCQYSV-----------TVVR 118
rgcckin    -------RRD-TIQRVRQCK----------YDKKKVILKDLKHCDG-----NFSEKQKIELNKLLQSDYYNLTKFYG-----------TVKLDTRIFGVV 104
humstakin  -------RRD-TIQRLRQCK----------YDKKRVILKDLKHNDG-----NFTEKQKIELNKLLQIDYYNLTKFYG-----------TVKLDTMIFGVI 107
pigstakin  -------RRD-TIQRLRQCK----------YDKKRVILKDLKHNDG-----NFTEKQKIELNKLLQIDYYNLTKFYG-----------TVKLDSMIFGVI 107
fgcckin    -------RRD-SGPRLQIRK----------YDKKVVILKDLKH-DE-----NFTEKQKMELNKLLQIDYYNLTKFYG-----------TVKLDNMIYAVI 102
rgcdkin    -----------QGEWVWLKKFEA------------GTAPDLRPSSLSL------------LRKMREMRHENVTAFLGLFVGPEVSA-----------MVL 126
rgcekin    -----------EGDWVWLKKFPG---------EHH---MAIRPATKMA------------FSKLRELRHENVALYLGLFLAGTADSPATPGEGILA-VVS 129
rgcfkin    -----------QGDWVWLKKFPP---------GDFGDIKSIKSSASDV------------FEMMKDLRHENVNPLLGFFYDSG----------MFA-IVS 130
olg3kin    -----------EGDWVWLKKIPI---------GK--TTTAVNQNTQNL------------FSQLREMRHENLNLYLGLFVDS----------GILALVVP 120
olg4kin    -----------EGDWVWLKKFEE---------GQF---KEVKQSTTKI------------FTKMKDLRNENVNPFLGFFLDCS----------MFA-VVT 125
olg5kin    -----------EGDLVWLKKCPT---------G---SVSSVSSSTEML------------FVKLRDMRHENLNLYLGLFFDS----------GIFG-IVT 129
cfy484kin  -----------EGDWVWLKKFPG---------DQH---IAIRPATKTA------------FSKLRELRHENVVLYLGLFLGSGGAGGSAAGEGVLA-VVS 132
cfyx83kin  -----------EGDWVWLKKFPG---------DQH---IAIRPATKTA------------FSKLRELRHENVVLYLGLFLGSGGAGGSAAGEGVLA-VVS 132
zk896kin   TYHRYKQRRPIK------------------------------------FIKEDMQLLTQMKQAVHDNLNPFLGAAFNEKEE----------MLVLW 112
gcy9kin    AIKRYQQVRNIT------------------------------------FPKSELRLLKELKICENDNLNKFYGISFNQQNE----------FIVMW 111
gcy4-kin   TIQYL------ENDLVLTTAHQ------V--------------------QELSQAEMMKFVKLRKLDHENLNKFIGLSIDGSRF-----------VSVW 115
gcy5-kin   TVQMF------EKDLVLTTKHH------S--------------------MQMNKEEKEKFVKLRKLEHDNLNKFIGLSIDGPQF-----------VAVW 112
gcy1-kin   TMMMY------EKEMVLTAKYQ------Y--------------------THLTKADKERFVKMRKLDHENINRFIGLSIDSAHF-----------ISVT 115
gcy2-kin   TMMMY------EKEMVLTAKYQ------Y--------------------THLTKADKERFVKMRKLDHENINRFIGLSIDSAHF-----------IAVT 113
gcy3-kin   EVKMF------EKDMVLTMKFQ------Y--------------------MNLNKADMDKFVKLRKLDHENLNKFIGLSIDSSQF-----------ISVT 111
gcy13kin   TQGYFVYMNE----PVLARKYQLRV-PIF--------------------KQDR-SELRM---LRSIEHDNVNRFIGLSIDGPVY-----------MSFW 119
f21h7kin   NFIFYHYHQEV----VAAKKHDLLFTVLF--------------------DANQKSEFRQ---MRNFDNDNLNKFIGLCLDGPQL-----------LSLW 114
zc412kin   NFIFYYYQQDI----LAAMKHDLILQ--F--------------------DAEQKAEFRQ---MRNFDNDNLNKFIGLCLDGPQL-----------FSLW 139
f52e1kin   NFLFFSLQRESDYEPVVAKKHA--YRPRL--------------------DDDKCTFMR---SLRNLDQDNLNRFIGLCLDGPQM-----------LSVW 120
gcy7-kin   NFLFFSLQRESDYEPVVAKKHA--YRPRL--------------------DDDKCTFMR---SLRNLDQDNLNRFIGLCLDGPQM-----------LSVW 118
b0024kin   NYLFFQIQNDVEMERVAAKKHS--IRMVF--------------------DNKTCATMRQVVEMRLIDHANLNKFIGMSLDAPQL-----------YSVW 123
gcy6-kin   NYLFFQIQNDVEMERVAAKKHS--IRMVF--------------------DNKTCATMRQ---MRLIDHANLNKFIGMSLDAPQL-----------YSVW 119
ruler      .......110.......120.......130.......140.......150.......160.......170.......180.......190.......200
```

Fig. 4

```
rgcakin     EYCPRG--SLQDILEN-----ESITLDWMFRYSLTNDIVK-------GMLFLHNGAICSHGNLKSSNCVVDGRFVLKITDYGLESFRDP----------- 190
rgcbkin     EYCPRG--SLQDILEN-----DSINLDWMFRYSLINDLVK-------GMAFLHNSIISSHGSLKSSNCVVDSRFVLKITDYGLASFRST----------- 191
egcbkin     EYCPRG--SLQDILEN-----ESINLDWMFRYSLINDIVK-------GMNFLHNSYIGSHGNLKSSNCVVDSRFVLKITDYGLASFRSS----------- 191
cionkin     EYCPKG--SLKDILQN-----DEIRLDWMFRFSLMNDIVK-------GMSFLHGSPIHSHGNLKSSNCVVDSRFVLKITDYGLSTFRSM----------- 253
gcx1kin     EYCPKG--SLEDILEN-----EKIELDKLMKYSLLHDLVK-------GLFFLHNSEIRSHGRLKSSNCVVDSRFVLKVTDFGLHRLHCL----------- 213
spgckin     HYCAKG--SLQDILEN-----DDIKLDSMFLSSLIADLVK-------GIVYLHSSEIKSHGHLKSSNCVVDNRWVLQITDYGLNEFK-K----------- 185
rgcgkin     QYCKKG--SLKDVLRN-----SDHEMDWIFKLSFVYDIVN-------GMLFLHGSPLRSHGNLKPSNCLVDSHMQLKLAGFGLWEFKHG----------- 201
drg0kin     --------SLEDVLAN-----EDLHLDHMFISSLVSDILK-------GMIYLHDSEIISHGNLRSSNCLIDSRWVCQISDFGLHELK-L----------- 68
drg1kin     DYCAKG--SLYDIIEN-----EDIKLDDLFIASLIHDLIK-------GMIYIHNSQLVYHGNLKSSNCVVTSRWMLQVTDFGLHELR-Q----------- 183
zc239kin    ELAQRG--SLKDILDN-----DDMPLDDVFRSQMTKDIIA-------GLEYLHSSPVGCHGRLKSTNCLIDARWMVRLSSFGLRELRGE----------- 242
gcy12kin    EYCSKG--SLHDILRN-----ENLKLDHMYVASFVDDLVK-------GMVYIHDSELKMHGNLKSTNCLITSRWTLQIADFGLRELREG----------- 193
rgcckin     EYCERG--SLREVLNDTISYPDGTFMDWEFKISVLNDIAK-------GMSYLHSSKIEVHGRLKSTNCVVDSRMVVKITDFGCNSILPP----------- 184
humstakin   EYCERG--SLREVLNDTISYPDGTFMDWEFKISVLYDIAK-------GMSYLHSSKTEVHGRLKSTNCVVDSRMVVKITDFGCNSILPP----------- 187
pigstakin   EYCERG--SLREVLNDTISYPDGTFMDWEFKISVLYDIAK-------GMSYLHSSKTEVHGRLKSTNCVVDSRMVVKITDFGCNSILAP----------- 187
fgcckin     EYCDKG--SLRDVLNDNISYPDGTFMDWEFKISVMYDIAK-------GMSYLHASKCEVHGHLKSTNCVVDGRMVVKITDFFGKSILSP----------- 182
rgcdkin     EHCARG--SLEDLLRN-----EDLRLDWTFKASLLLDLIR-------GLRYLHHRHF-PHGRLKSRNCVVDTRFVLKITDHGYAEF-------------- 197
rgcekin     EHCARG--SLHDLLAQ-----RDIKLDWMFKSSLLLDLIK-------GMRYLHHRGV-AHGRLKSRNCVVDGRFVLKVTDHGHGRL-------------- 200
rgcfkin     EFCSRR--SLEDILTQ-----DDVKLDWMFKSSLLLDLIK-------GMKYLHHREF-IHGRLKSRNCVVDGRFVLKVTDYGFNNI-------------- 201
olg3kin     EHCPRG--SLADLLAD-----SDVRLDWMFKSSLLMDLIK-------GMKYLHLRGL-THGRLKSTNCLVDGRFVLKITDYGLPMI-------------- 191
olg4kin     EHCSRG--SLQDLLRN-----EDVKLDWMFKSSLLLDLIK-------GMKYLHHREF-PHGRLKSRNCVVDGRFVLKITDYGFNEL-------------- 196
olg5kin     EHCIRG--SLEDLLAN-----EDVRLDWMFKSSLLMDLIR-------GMKYLRHRNI-IHGRLKSRNCVVDGRFVLKVTDYGFNEI-------------- 200
cfy484kin   EHCARG--SLHDLLAQ-----RDIKLDWMFKSSLLLDLIK-------GMRYLHHRGV-AHGRLKSRNCVVDGRFVLKVTDHGHARL-------------- 203
cfyx83kin   EHCARG--SLHDLLAQ-----RDIKLDWMFKSSLLLDLIK-------GMRYLHHRGV-AHGRLKSRNCVVDGRFVLKVTDHGHARL-------------- 203
zk896kin    KFCSRGFTTIQDII-----YNANVVLDEKFHGAFVRDITL-------GLEYLHASPIGYHGSLTPWCCLIDRNWMVKLS-F-TDYGIANPLERWEKQGAI 198
gcy9kin     VLCSRG--SLEDIL-----FNDELKLGRNFQVSFAKDVVK-------GLNFLHTSPLLHHGMLCLQNCLVDSNWTVKLTNFATEAVIFEKLDHNELRPFI 197
gcy4-kin    KMCSRG--SLQDIISKG---SFSMDY--FFMFCMIRDIAE-------GLHYIHKSFLRHHGNLRSATCLVNDSWQVKLADFGLQFLQDE----------- 190
gcy5-kin    KMCSRG--SLQDIIARG---NFSMDG--FFMFCIITDIAE-------GMNFLHKSFLHLHGNLRSATCLVNDSWQVKLTDFGLGALL-E----------- 186
gcy1-kin    KLCSRG--SLQDILSRG---NFSMDY--FFMFCIIRDVAK-------GLEYLHKTFLRLHGNLRSATCLVNDSWQVKLAEYGMDNLV-E----------- 189
gcy2-kin    KLCSRG--SLQDILSRG---NFSMDY--FFMFCIIRDVAK-------GLEYLHKTFLRLHGNLRSATCLVNDSWQVKLAEYGMDNLV-E----------- 187
gcy3-kin    KLCSRG--SLLDILYKG---NFSMDF--FFMYCIIKDVAE-------GMSYLHKSFLRLHGNLRSATCLVNDSWQVKLAEFGFDQLL-E----------- 185
gcy13kin    RYCSRG--SIKDVIAKS---SINMD--GFFIYCLIKDIASVSRHENPGLQYIHHSPIKQHGSLTSECCYINDRWQVKIGSYGLSFMQGV----------- 201
f21h7kin    RFCSRG--SLSDVISKS---SMQMDFTSFFMFSLIRDISN-------GLYFIHSSFLKCHGQLTSRCCLIDDRWQIKISGFGLKSVRTF----------- 191
zc412kin    RFCSRG--SLSDVISKS---SMQMD--SFFMFSLIRDISN-------GLLFIHNSFLKCHGHLTSRCCLIDDRWQIKISGYGLKSVRTF----------- 214
f52e1kin    RFCSRG--SIADVILKA---TIQMD--NFFIYSLIKDMVH-------GLVFLHGSMVGYHGMLTSKCCLIDDRWQVKISNYGLQDLRSP----------- 195
gcy7-kin    RFCSRG--SIADVILKA---TIQMD--NFFIYSLIKDMVH-------GLVFLHGSMVGYHGMLTSKCCLIDDRWQVKISNYGLQDLRSP----------- 193
b0024kin    RFCSRG--SLADVIRKA---SMQMD--GFFIYSLMKDIIN-------GLTWIHESSHEFHGMLTSKNCLLNDRWQLKITDFGLRIFRTH----------- 198
gcy6-kin    RFCSRG--SLADVIRKA---SMQMD--GFFIYSLMKDIIN-------GLTWIHESSHEFHGMLTSKNCLLNDRWQLKITDFGLRIFRTH----------- 194
ruler       .......210.......220.......230.......240.......250.......260.......270.......280.......290.......300
```

Fig. 4

```
rgcakin    ---E---PE----QGHTLFAKKLWTAPELLR-MAS-P--PARGS-----------QAGDVYSFGIILQEIALRSGVFYV----EGLD---------LSPK 252
rgcbkin    --AE---PD----DSHALYAKKLWTAPELLS-GNP-L--PTTGM-----------QKADVYSFAIILQEIALRSGPFYL----EGLD---------LSPK 254
egcbkin    --CE---NE----DSHALYAKKLWTAPELLI-YDR-H--PPQGT-----------QKGDVYSFGIILQEIALRNGPFYV----DGMD---------LSPK 254
cionkin    --SK---YE----DSDNFYEKKLWTSPELLR-SPI-P--PPNGS-----------QKGDVYSFAIIVHEVALRKGTFYV----GGIH---------FGGR 316
gcx1kin    --EE---INLEEIGEHAYYKKMLWTAPELLR-DSN-A--PPMGT-----------QKGDIYSFAIILHEMMFRKGVFAL----ENED---------LSPN 280
spgckin    --GQ---KQDVDLGDHAKLARQLWTSPEHLR-QEGSM--PTAGS-----------PQGDIYSFAIILTELYSRQEPF------HENE---------MDLA 251
rgcgkin    --STCRIYNQEATDHSELY----WTAPELLR-LRE-L--PWSGT-----------PQGDVYSFAILLRDLIHQQAHGPF----EDLE---------AAPE 267
drg0kin    --GQ---EEPNK--SELELKRALCMAPELLR--DAYR--PGRGS-----------QKGDVYSFGILLYEMIGRKGPW------GDTA---------YSKE 131
drg1kin    --CA----ENESIGEHQHYRNQLWRAPELLR-NH------IHGS-----------QKGDVYAFAIIMYEIFSRKGPF------GQIN---------FEPK 244
zc239kin   -------ETWQQEDDVQEGKDQLWTSPELLRWSTGLSQCGVLLV-----------QKSDVYSLAIVLYELFGRLGP------WGD---------EPMEPR 309
gcy12kin   -------IMYDSSYNIWE--NFLWTAPEAMTINGSLAISNPP-T-----------PKADAYSFGIIFHEIFTREGPYKIYVQRGDVNGEAAPKKDSVECR 272
rgcckin    -------------------KKDLWTAPEHLR--------QATIS-----------QKGDVYSFSIIAQEIILRKETFYTLSCRDQ--------------N 232
humstakin  -------------------KKDLWTAPEHLR--------QANIS-----------QKGDVYSYGIIAQEIILRKETFYTLSCRDR--------------N 235
pigstakin  -------------------KKDLWTAPEHLR--------RASVS-----------QKGDVYSYGIIAQEIILRRETFYTLSCRDQ--------------K 235
fgcckin    -------------------EKDLWTAPEHLH--------QEGFS-----------QKGDVYSYGVIAQEIILRQETFYTEQCDDT--------------K 230
rgcdkin    ------LESHCSFRPQPAPEELLWTAPELLR-GPRGPWGPGKAT-----------FKGDVFSLGIILQEVLTRDPPYCS------WG---------LSAE 264
rgcekin    ------LEAQRVLPEPPSAEDQLWTAPELLR-DPALER---RGT-----------LAGDVFSLGIIMQEVVCRSTPYAM------LE---------LTPE 264
rgcfkin    ------LEMLRLSEEEPSEEELLWTAPELLR-APGGIR---LGS-----------FAGDVYSFAIIMQEVMVRGAPFCM------MD---------LSAK 265
olg3kin    ------LQSQNLSLPED-PQDLLWTFPELLR-NPV--R---EGS-----------FAGDVFSFSIIIQEVISRTLPYAM------MD---------MPAH 252
olg4kin    ------LESQKAPVEEPPPEELYWTAPELLR-DLTLFH---KGT-----------YKGDVYSFSIILQEVVVRGPPYCM------LG---------LPPE 260
olg5kin    ------MNTQEVDVEEEKAEDLLWTAPELLR-NLNLRR---KGS-----------FPGDVYSFAIIMQEVVSRSAPFCM------LD---------MPPK 264
cfy484kin  ------MEAQRVLLEPPSAEDQLWTAPELLR-DPALER---RGT-----------LPGDVFSLGIIMQEVVCRSAPYAM------LE---------LTPE 267
cfyx83kin  ------MEAQRVLLEPPSAEDQLWTAPELLR-DPALER---RGT-----------LPGDVFSLGIIMQEVVCRSAPYAM------LE---------LTPE 267
zk896kin   EI--AAAKDSDDKSQASQATSIIYMAPELLKNRETNKRRGMDQSFTWVKQSMLRRQAGDIYSFGMVMYEILFRSLPF---------------RD-NTNIS 280
gcy9kin    NTDSESADDVSDPTKDFARKKYLQQAPEII--REIVTTKTIPEG----------SQSADIYALGMVLYQILFRVEPF---------------HERNKSIN 270
gcy4-kin   --EEKPFKK----------N-MLWAAPEVIRGSLSIEQ----MD-----------SSADIYSFAIVASEILTKREAWDLSRRKE---------------- 246
gcy5-kin   --EHTPSKK----------R-LLWAAPEVLRGSLTIHQ----MD-----------PSADVYSFAIIASEILTKREAWDISNRKEG-------------AD 245
gcy1-kin   --EQTPPKK----------R-LLWVAPEVLRGSLSVSQ----ME-----------PSADIYSFAIIASEILTKKEAWDILDRKED-------------CE 248
gcy2-kin   --EQTPPKK----------R-LLWVAPEVLRGSLSVSQ----ME-----------PSADIYSFAIIASEILTKKEAWDILDRKED-------------CE 246
gcy3-kin   --EITPTKR----------R-LLWAAPEVLRGSLTVSQ----MD-----------PSADVFSFAIIASEILTRKEAWDLKERKEG-------------YD 244
gcy13kin   --EKR--TE----------DGLLHTAPEVLREGLTSG------T-----------QAGDVYSFSIVCSELVGHSSAWNLENRKEE-------------AD 257
f21h7kin   --ENPKFTK----------EDLLWASPEYLRNEDQER------L-----------PEGDIYSFGIICAEILTRSSAFDLENRKEK-------------PD 249
zc412kin   --ENPK--K----------EDLLWTPPENLRNENEER------L-----------PEGDIYSFGIICSEILTRSSAFDLENRKEK-------------PD 270
f52e1kin   --E--MYEK----------KDLLWSAPELLRAEDIKG------S-----------KEGDVYSLGIICAELITRKGVFNMEDRKED-------------PE 251
gcy7-kin   --E--MYEK----------KDLLWSAPELLRAEDIKG------S-----------KEGDVYSLGIICAELITRKGVFNMEDRKED-------------PE 249
b0024kin   --D--QYNK----------SDRLWTSPELLRTDDILG------S-----------REGDIYSFGIISAELITRSSVFDLENRKED-------------AE 254
gcy6-kin   --D--QYNK----------SDRLWTSPELLRTDDILG------S-----------REGDIYSFGIISAELITRSSVFDLENRKED-------------AE 250
ruler      .......310.......320.......330.......340.......350.......360.......370.......380.......390.......400
```

Fig. 4

```
rgcakin     EIIERVTR---GEQPPFRPSMDLQ-------SHLEELGQLMQRCWAEDPQERPPFQQIRLAL-RKFNKENSS------ 313
rgcbkin     EIVQKVRN---GQRPYFRPSIDRT-------QLNEELVLLMERCWAQDPTERPDFGQIKGFI-RRFNKEGGT------ 315
egcbkin     EIVQKVRN---GQKPYFRPTTDTS-------CHSEELSILMEGCWAEDPADRPDFSYIKIFV-MKLNKEGSTSIL--- 318
cionkin     GIITKVRN---GCYPYFRPLLDHS-------ILSDDLCLLMQRCWAEEPSERPDFQQLKDII-KKFNKENGG------ 377
gcx1kin     EIVQRVRKPVSEDQEPLRPWVSETGEGEGDDALNDTLLSLMVACWSEDPHERPEVSSVRKAV-RSLNRDNETS----- 352
spgckin     DIIGRVKSGEV---PPYRPILNAV--NAA---APDCVLSAIRACWPEDPADRPNIMAVRTML-APLQKGLKP------ 314
rgcgkin     EIISCIKDPRAPV--PLRPSLL-------EDKGDERIVALVRACWAESPEQRPAFPSIKKTL-RE---ASPRGRV--- 329
drg0kin     EIIQFVKCPEMLQHGVFRPALTHT--HLD---IPDYIRKCLCQCWDEDPEVRPDIRLVRMHL-KELQAGLKP------ 197
drg1kin     EIVDYVKKLPLKGEDPFRPEVESIIEAES---CPDYVLACIRDCWAEDPEERPEFSVIRNRL-KKMRGGKTK------ 312
zc239kin    EIVSLVKREALAGKKPFRPDMAVLKE------SPRIVQETVVAAWTEDPLNRPSLHQIKRKL-KPL--TIGL------ 372
gcy12kin    ALVEKTVRRVYSDPY-FRPDTSDL-E------VQNYVKEVMAACWHHDPYQRPEFKTIKNKL-KPL--FHQIYKQ--- 336
rgcckin     EKIFRVENSY--GTKPFRPDLFL----ETADEKELEVYLLVKSCWEEDPEKRPDFKKIESTLAKIFGLFHDQKNE--- 301
humstakin   EKIFRVENSN--GMKPFRPDLFL----ETAEEKELEVYLLVKNCWEEDPEKRPDFKKIETTLAKIFGLFHDQK----- 302
pigstakin   EKIFRVENSN--GVKPFRPDLFL----ETAEEKELEVYLLVKNCWEEDPEKRPDFKKIENTLAKIFGLFHDQK----- 302
fgcckin     EKISRVQNNK--GECPFRPDLNL----DSANEREIEVYVLVKSCWEEDPERRPDFKKIENTLSKIFSNFHSQTTE--- 299
rgcdkin     EIIRKVASPP----PLCRPLVSP-------DQGPLECIQLMQLCWEEAPDDRPSLDQIYTQF-K------------- 316
rgcekin     EVIQRVRSPP----PLCRPLVSM-------DQAPMECIQLMAQCWAEHPELRPSMDLTFDLF-KGINKGRKT------ 324
rgcfkin     EVIDRLKMPP----PVYRPVVSP-------EFAPPECLQLMKQCWAEAAEQRPTFDEIFNQF-KTFNKGKKT------ 325
olg3kin     EIVDRLMKPP----PLCRPVVSV-------DEAPAECLNLMNECWNEDPTKRPTFDDIFKQF-RGISRGRRA------ 312
olg4kin     EIIRKVKKPP----PMCRPTVAP-------DQAPLECIQLMKQCWSEQPDRRPTFEEIFDRF-KIINKGKKT------ 320
olg5kin     EIISKVKESP----PLCRPVVSV-------EEAPLDIIQLMKQAWSEEPDKRPTFEEIFKQF-KGNSKAKKT------ 324
cfy484kin   EVVERVRSPP----PLCRPSVSM-------DQAPVECIQLMKQCWAEHPDLRPSLGHIFDQF-KSINKGRKT------ 327
cfyx83kin   EVVERVRSPP----PLCRPSVSM-------DQAPVECIQLMKQCWAEHPDLRPSLGHIFDQF-KSINKGRKT------ 327
zk896kin    ELVDYLA---DGSK-TVSPEI---QN-QMFTGLHPDLNALLRDCWSENPEIRPSIRRVRLNTEMVLKT---------- 340
gcy9kin     KLMETLAMANDDDQ-LIRPTF---PSSNTGEGYNLQLLSCIEACWLEIPEMRPPIKKVR----TMVNA---------- 330
gcy4-kin    EIKYAVK--K-GGQFVLRPDL----HIDI--EVNQTLLALVKDCWCENPEERPSAENVCKVLFDMTPNTED------- 308
gcy5-kin    EILYMVK--K-GGNRTIRPEL----ILDA--EVSPRLTTLVKDCWSEQPEDRPKAEQICKLLSEMTPRGNT------- 307
gcy1-kin    EIVYNVK--K-GGLFPIRPEI----ITDIH-DVNPALIALVKDCWAEVPEDRPTAENICSQMKGLVSKQKT------- 311
gcy2-kin    ---------------------------------ALIALVKDCWAEVPEDRPTAENICSQMKGLVSKQKT------- 282
gcy3-kin    EIIYRVK--K-GGSFPIRPDI----ITDVP-DVNPTLIALVKDCWAEAPEDRPTAENICEQLRDLMPKTKS------- 307
gcy13kin    EIIFMVK--R-GGRTPFRPSL----DDVDD-DINPAMLHLIRDCWDEDPKQRP------------------------ 302
f21h7kin    VIIYQVKFTK-GGHNPTRPSL----DTGETVVINPALLHLVRDCWTERPSERPSIEQVRSHLNGMKDGRKT------- 315
zc412kin    VIIYQVK--K-GGHNPMRPSL----DTGETVEINPALLHLIRDCWTERPSERPSIEQVRGHLNGMRDGRKS------- 334
f52e1kin    EIIYLLK--K-GGLKSPRPDL----EYDHTIEINPALLHLVRDCFTERPSERPSIETVRSQLRGMNSSRNDNLMDHVF 322
gcy7-kin    EIIYLLK--K-GGLKSPRPDL----EYDHTIEINPALLHLVRDCFTERPSERPSIETVRSQLRGMNSSRND------- 313
b0024kin    EIIYMLK--K-GGLQSPRPSL----EHDESIEINPALLHLVRDCWTERPSERPDIKQVASQLRSMNTNRND------- 318
gcy6-kin    EIIYMLK--K-GGLQSPRPSL----EHDESIEINPALLHLVRDCWTERPSERPDIKQVASQLRSMNTNRND------- 314
ruler       .......410.......420.......430.......440.......450.......460.......470........
```

SPECIES	CLONE	PREDICTED SEQUENCE
Rat	GC-A	AICS**HGN**LKSSNCVVDGRFVLKITDYG
Rat	GC-B	IISS**HGS**LKSSNCVVDSRFVLKITDYG
Rat	GC-E	RGVA**HGR**LKSRNCVVDGRFVLKVTDFG
C. elegans	GCY-X_1	EIRS**HGR**LKSSNCVVDSRFVLKVTDF-
Drosophila	DGC-1	QLVY**HGN**LKSSNCVVTSRWMLQVTDFG
Drosophila	DMGCR	EIIS**HGN**LRSSNCLIDSRWVCQISDFG
Rat	JAK2	KSLI**HGN**VCAKNILLIREEDRKTGNPP
Human	TYK2	KNLV**HGN**VCGRDNILLARLGLAEGTSP
Arabidopsis	TMKL1	VPII**HGN**IRSKNVLVDDFFFARLTEFG
Arabidopsis	ATKLD3	GTTS**HGN**IKSSNILLSDSYEAKVS---
Arabidopsis	EST	----**HGN**LXSKNXLLDXXFRPRVSDFG
Arabidopsis	EST	---T**HGN**LKSSNVLLGPDFESCLTDYG
PROTEIN TYROSINE KINASE CONSENSUS		----HRDL---N-----K------DFG

Fig. 5. Alignment of signature sequences of protein-kinase homology domains from various proteins. Sequences encoding protein-kinase homology domains from transmembrane guanylyl cyclases, the JAK family of protein kinases, and members of a family of receptor-like molecules identified in *A. thaliana* are compared with a protein tyrosine kinase consensus sequence

tide-dependent regulation (Goraczniak et al. 1992, Koller et al. 1993). (3) In the presence of Mg^{2+}, ATP fails to compete with GTP for formation of cyclic GMP. Thus, the catalytic site does not represent a site of ATP binding in the presence of Mg^{2+}. (4) The KHD is strongly conserved across species and receptor subtypes suggesting significant evolutionary pressure for conservation of the domain (Fig. 4). Since many of the amino acids are also conserved within the catalytic domain of protein kinases, the required binding of ATP or other molecules, such as regulatory proteins, likely represents the basis of the pressure.

The membrane forms of guanylyl cyclase are analagous in some respects to signaling pathways where hormone binding to a 7-transmembrane domain receptor activates a guanine nucleotide binding protein, subsequently

Fig. 4. Sequence alignment of protein-kinase homology domains. Sequences encoding protein-kinase homology domains from transmembrane guanylyl cyclases were aligned to identify signature sequences of protein-kinase homology domains. Amino acids in GC-A that have been identified as phosphorylated are colored red (Potter and Hunter 1998)

leading to the activation of an adenylyl cyclase. However, we have yet to find ATPase activity associated with the purified GC-A receptor (unpublished data). The caveat with respect to these experiments is the possible existence of ATPase activating proteins that are absent in the purified preparations.

In the case of GC-A, GC-B and the sea urchin sperm guanylyl cyclases, it seems clear that the receptors exist in a phosphorylated state in the absence of added ligand (Garbers and Lowe 1994). The addition of ligand to cells causes a rapid dephosphorylation of each of these receptors and this is correlated with desensitization (Bentley et al. 1986, Potter and Garbers 1992). Amino acids in GC-A that are phosphorylated in the basal state have been identified, and single mutation of these amino acids to alanine reduces ligand stimulated guanylyl cyclase activity. Ser497, Thr500, Ser502, Ser506, Ser510, and Thr513 are sites phosphorylated in GC-A and mutagenesis of all of the sites to alanine results in a complete loss of ligand-stimulated GC-A activity (Potter and Hunter 1998). GC-B, which is desensitized in a manner similar to GC-A, has serine and threonine residues conserved with those identified as GC-A phosphorylation sites (Potter 1998). Dephosphorylation of natriuretic peptide receptor guanylyl cyclases, therefore, likely represents one mechanism regulating desensitization (Potter and Garbers 1992, Potter 1998). Consistent with phosphorylation modulating the ligand-sensitivity of natriuretic peptide receptor guanylyl cyclases, the *in vitro* sensitization of GC-A has recently been accomplished. In a preparation of crude membranes, an initial incubation with the protein kinase substrates ATP or ATPγS and protein phosphatase inhibitors such as microcystin renders GC-A sensitive to stimulation by atrial natriuretic peptide (ANP) and a non-hydrolyzable analog of ATP, AMPPNP. Sensitization of GC-A was reversible and correlated well with phosphorylation of GC-A (Foster and Garbers, 1998). Thus, the phosphorylation state of natriuretic peptide receptor guanylyl cyclases likely represents a mechanism to regulate their sensitivity to ligand (Fig. 6). The possibility that other proteins exist whose state of phosphorylation regulates GC-A activity also remains possible. Unfortunately, the regulatory protein phosphatases and protein kinases have not been identified.

It is intriguing that the phosphorylation sites identified in GC-A are localized to a region in the N-terminus of the KHD (Fig. 4). The homologous region in other receptor guanylyl cyclases also contains numerous serine and threonine residues (Fig. 4). In addition, in a recently cloned *C. elegans* receptor guanylyl cyclase (GCY-X1), a 150-amino acid insertion containing 8 consensus cAMP-dependent protein kinase phosphorylation sites occurs in this region (not shown in Fig. 4) (Baude et al. 1997). Thus, in many different cyclases this region may contribute to regulation of activity.

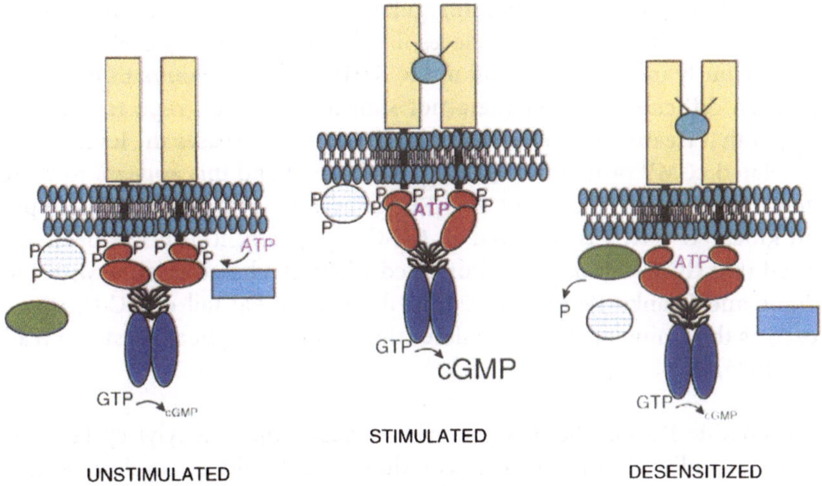

Fig. 6. Model for regulation of transmembrane guanylyl cyclases. The transmembrane guanylyl cyclase is phosphorylated in the basal state and phosphorylation is likely required for maximal ligand-dependent guanylyl cyclase activity. There may also be other proteins whose phosphorylation state regulates the activity of guanylyl cyclases (green checked circle). A guanylyl cyclase protein kinase (light blue rectangle) phosphorylates the guanylyl cyclase rendering it sensitive to ligand. A guanylyl cyclase phosphoprotein phosphatase (green ellipse) is also present and is either activated in response to ligand, or preferentially recognizes the liganded form of the phosphorylated guanylyl cyclase as a substrate. Phosphates are represented by "P". ATP serves as a substrate for phosphorylation of the guanylyl cyclase and facilitates signal transduction by binding to a regulatory site. The activities of the GC-kinase and GC-phosphatase regulate the ligand-dependent activity of transmembrane guanylyl cyclase

A recent report demonstrated that a partially purified preparation of photoreceptor guanylyl cyclase was capable of being phosphorylated *in vitro*, however, the significance of this phosphorylation is not clear. There were no reported changes in enzyme activity that correlated with changes in phosphorylation state (Aparicio and Applebury 1996). One possible explanation for this is that phosphorylation affects only ligand-dependent guanylyl cyclase activity, and in the absence of known ligands for the photoreceptor cyclases this can not be examined. The two known photoreceptor guanylyl cyclases have serines that correspond to a known phosphorylation site of GC-A and GC-B, thus it is possible that a similar mechanism of regulation exists for these enzymes (Fig. 6). Indeed, understanding the role of phosphorylation in regulating photoreceptor guanylyl cyclases may shed new light on current models of phototransduction.

The role of phosphorylation in regulating the STa receptor (GC-C) is not clear. GC-C is one of the few cyclases that is not particularly Ser/Thr rich in the previously mentioned region of the KHD, and phosphorylation of GC-C has been difficult to detect (data not shown). However, reports exist that suggest that treatment of cells with phorbol esters increases the level of STa-stimulated cGMP production (Weikel et al. 1990) and this appears to correlate with increased ^{32}P content of GC-C. Consistent with a model where protein kinase C (PKC) is involved in GC-C regulation, it also has been suggested that GC-C can be phosphorylated when incubated *in vitro* with PKC (Crane and Shanks 1996). Ser1029, in the C-terminal tail of GC-C, is proposed as the amino acid that regulates the response to phorbol esters (Wada et al. 1996).

Amphipathic Region. Soluble guanylyl cyclases, like adenylyl cyclases, require heterodimerization of catalytic domain subunits for cyclase activity, while transmembrane guanylyl cyclases are active as homomers. While both subunits of the adenylyl cyclase catalytic domain dimer are found within one molecule, guanylyl cyclases must oligomerize to form the dimeric species. Apparently dimerization is not induced by ligands since GC-A exists as a higher-ordered structure even in the absence of ANP (Chinkers and Wilson 1992, Lowe 1992), and photoreceptor guanylyl cyclases (GC-E and GC-F) migrate as dimers on gel filtration columns in the absence of known ligands (Yang and Garbers 1997). Soluble guanylyl cyclase subunits can be co-purified indicating that they too are dimeric in nature (Garbers 1979, Kamisaki et al. 1986, Humbert et al. 1990).

Amino-terminal to the cyclase homology domain lies a stretch of amino acids predicted to form an amphipathic α-helix. Also known as a leucine-zipper, this motif is involved in dimerization of several molecules including Fos and Jun (Landschulz et al. 1988). Thus, it was predicted that this domain would contribute to dimerization of guanylyl cyclases. Indeed, Chinkers and Wilson (1992) demonstrated that GC-A deletion mutants lacking portions of the predicted amphipathic domain no longer migrated as dimers on gel filtration columns, and no longer retained guanylyl cyclase activity. In contrast, deletion of the majority of the amino-terminus of the soluble cyclase subunits while retaining the leucine zipper region results in molecules that continue to dimerize and retain enzymatic activity (Wedel et al. 1995). Thus, the putative amphipathic α-helical region likely serves as a dimerization domain.

Although the amphipathic region has been suggested as both necessary and sufficient for dimerization of guanylyl cyclases, sequences outside this domain may also contribute to dimer formation. Other portions of the molecules are likely involved in dimerization since even though heterodimerization of GC-A and GC-B has been obtained when over-expressed in tissue culture (Chinkers and Wilson 1992), a native heteromer has not been reported despite both GC-A and GC-B containing nearly identical sequences within their amphipathic region (Fig. 7). Similarly, the photoreceptor guanylyl cyclases (GC-E and GC-F) preferentially form homomers in the ROS even though they are expressed in the same cell and contain nearly identical amphipathic regions (Yang and Garbers 1997). In fact, the high degree of sequence conservation in this domain among numerous guanylyl cyclases is quite striking and suggests a strong pressure to retain these sequences (Fig. 7). Recent studies on the genetic basis of a cone-rod dystrophy that maps to human chromosome 17 support an important functional role for the amphipathic region in that various point mutations within the amphipathic region of GC-E were associated with this retinal degeneration (Kelsell et al. 1998, Perrault et al. 1998).

Heme Binding Domain. Soluble guanylyl cyclases are primary targets of the signaling molecule NO (Snyder 1992). A heterodimer of α- and β-subunits, the soluble enzyme is stimulated by binding of NO to a prosthetic heme group. Although sequences predicted to encode soluble guanylyl cyclases have been found in a variety of species, two isoforms each of α- and β-subunits have been identified in mammals (Hobbs 1997). Co-expression of either of the α-subunits with β_1 is required for detectable guanylyl cyclase activity although only the α_1-β_1 heterodimer has been identified in tissues. An enzymatically active complex with β_2 has not been identified.

The soluble enzymes share several topological characteristics including a C-terminal guanylyl cyclase catalytic domain, which is preceded by a predicted amphipathic region that likely participates in dimerization (see above) (Fig. 1). Co-expression of deletion constructs of α-and β-subunits lacking the amino-terminus results in proteins that form stable dimers and are enzymatically active, but are no longer NO stimulated (Wedel et al. 1995). The N-terminus of the molecules has been demonstrated to bind heme, although the precise localization and number of the heme binding site(s) is controversial. Initially purified in the absence of heme (Garbers 1979), the soluble guanylyl cyclase is now known to require this prosthetic group for NO-stimulated enzyme activity (Gerzer et al. 1981).

Estimates of the stoichiometry of heme binding have varied, with initial reports suggesting that one molecule of heme bound to the heterodimeric enzyme (Gerzer et al. 1981). Stone and Marletta (1995a) subsequently purified the soluble cyclase containing 1.5 equivalents of heme per heterodimer

```
olgahin      ---LVGEQAKAQDGLKKRLGKAKAALENAHQALEEEKKKTVDLLFSI-------- 44
dmu117hin    -VILVGEQARAQDGLRRRMDKIKNSIEEANSAVTKERKKNVSLLHLI-------- 46
hsgcsahin    -----------QDGLKKRLGKLKATLEQAHQALEEEKKKTVDLLCSIFPCEVAQQ 44
rga1hin      -V-LIGEQARAQDGLKKRLGKLKATLEHAHQALEEEKKKTVDLLCSI-------- 45
rgb1hin      -LVLLGEQFREEYKLTQELEILTDRLQLTLRALEDEKKKTDTLLYSV-------- 46
bsb1hin      -LVLLGEQFREEYKLTQELEILTDRLQLTLRALEDEKKKTDTLLYSV-------- 46
olgbhin      --VLLGEQFREEYKLTQELEILTDRLQHTLRALEDEKKKTDRLLYSV-------- 45
rgb2hin      -LILLNQQRLAEMELSCQLEKKKEELRVLSNHLAIEKKKTETLLYAM-------- 46
hsgcsbhin    --VLLGEQFREEYKLTQELEILTDRLQLTLRALEDEKKKTDTLLYSV-------- 45
b0024hin     NL--MDHVFNVLESYASTLEDE---VAERMKELVEEKKKSDVLLYR--------M 42
gcy31hin     S----AELKLLLHQEAQKSRNM----RENMNRLKKERRRTDKLLYQ--------M 39
gcy32hin     -LSDV------E--VNLQLEANNEQLETMTRELELERQKTDSILKD--------M 38
gcy33hin     S----DTLKTMLENEKRRSEVL----TEMTREISEAKKTARTLLTQ--------M 39
t04d3hin     -MSQV------ELNRTL--EETTKKLKKMAQELEIEKQKTDELLCEF------TL 40
m04g12hin    -LSDV------EFTVNLQLEANNEQLETMTHELEVERQKTDSILKD--------M 40
dmu123hin    S---VANELRHQRPVPPKRYDSVTLMFSGIVGFGQYCAANTDPDGAM---KIVKM 49
b0240hin     --SLVDLMIKNLTAYTQGLNE--T-VKNRTAELEKEQEKGDQLLME--------L 42
c30g4hin     S-----------------QYEDSGPVGGATMNF--Y--TNGIRKAAI---RR-NQ 30
c54e4hin     -ICQM------ELNKKL--EETMKKMKKMTEELEVKKSQTDRLL--F------EF 38
cfy484hi     N--IIDSMLRMLEQYSSNLEDL---IRERTEELELEKQKTDRLLTQ--------M 42
cfyx83hi     N--IIDSMLRMLEQYSSNLEDL---IRERTEELELEKQKTDRLLTQ--------M 42
drg0hin      NI--FDNMLSIMEKYAYNLEGL---VQEANQFVVRREEETDMLLYQ--------M 42
drg1hin      N--IMDQMMEMMEKYANNLEDI---VTERTRLLCEEKMKTEDLLHR--------M 42
egcbhin      N-----NLLSRMEQYANNLENL---VEERTQAYLEEKRKAENLLYQ--------I 39
f21h7hin     NLFTMDHVFNMLETYASTLEEE---VSDRTKELTEEKKKSDVLLYR--------M 44
f52e1hin     N---------MLESYASSLEEE---VSERTKELVEEKKKSDVLLYR--------M 35
fgcchin      --SYMDTLIRRLQLYSKNLEHL---VEERTQLYKAERDRADRLNYL--------L 42
gc9hin       KGSLVDQMMKMMEEYTANLENM---VRDRTALLEEAQKQADRLLNS--------M 44
gcxhin       NL--VDNLLKRMEQYANNLEGL---VEERTQEYLAEKKKVEELLHQ--------L 42
gcy10hin     --NLVDQMIRMSEKYADELEQM---VAIRTADLADAQMQTMRLLNE--------M 42
gcy12hin     N--IMDHMVLMMEKYQTQLEDL---VDERTIELKDEQRRSQHLLQR--------M 42
gcy13hin     NL--MDHVFSVLEKHASSLEDE---VQERMKELVEEKKKSDILLYR--------M 42
gcy1hin      NL--MDHVFNMLEEYTSTLEEE---IEERTKELTLEKKKADILLSR--------M 42
gcy2hin      NL--MDHVFNMLEEYTSTLEEE---IEERTKELTLEKKKADILLSR--------M 42
gcy3hin      NL--MDHVFNMLEEYTSTLEVE---VEERTKELTLEKKKADLLLSR--------M 42
gcy4hin      NL--MDHVFSMLEEYTSSLEVE---VGERTKELTLEKKKSDILLGR--------M 42
gcy5hin      NL--MDHVFNMLEEYTSTLEVD---IEERTKELTLEKKKADILLSR--------M 42
gcy6hin      NL--MDHVFNVLESYASTLEDE---VAERMKELVEEKKKSDVLLYR--------M 42
gcy7hin      NL--MDHVFNMLESYASSLEEE---VSERTKELVEEKKKSDVLLYR--------M 42
gcy9hin      --SLVDQMMKMMEEYTANLENM---VRDRTALLEEAQKQADRLLNS--------M 42
cionhin      NL--VENLLHRMEQYANNLEGL---VEERTAAYLEEKRKADELLYQ--------M 42
humstahin    NESYMDTLIRRLQLYSRNLEHL---VEERTQLYKAERDRADRLNFM--------L 44
olg3hin      N--IIDSMLRMLEQYSSNLEDL---IRERTDELEVERNKTEKLVGQ--------L 42
olg4hin      N--IIDSMLRMLEQYSSNLEDL---IRERTEELEVEKQRTEKLLSE--------M 42
olg5hin      N--IIDSMLRMLEQYSSNLEDL---IRERTEELEVERQKTDNLVAQ--------M 42
pigstahin    NESYMDTLIRRLQLYSRNLEHL---VEERTQLYKAERDRADRLNFM--------L 44
rgcahin      NI--LDNLLSRMEQYANNLEEL---VEERTQAYLEEKRKAEALLYQ--------I 42
rgcbhin      SI--LDNLLLRMEQYANNLEKL---VEERTQAYLEEKRKAEALLYQ--------I 42
rgcchin      --SYMDTLIRRLQLYSRNLEHL---VEERTQLYKAERDRADHLNFM--------L 42
rgcdhin      S--VADSMLRMLEKYSQSLEGL---VQERTEELELERRKTERLLSQ--------M 42
rgcehin      N--IIDSMLRMLEQYSSNLEDL---IRERTEELEQEKQKTDRLLTQ--------M 42
rgcfhin      N--IIDSMLRMLEQYSSNLEDL---IRERTEELEIEKQKTEKLLTQ--------M 42
rgcghin      --SILDSMMGKLEMYASHLEEV---VEERTCQLVAEKRKVEKLLST--------M 42
spgchin      N--ILDNMIAIMERYTNNLEEL---VDERTQELQKEKTKTEQLLHR--------M 42
zc239hin     KRTIMDNMVSMIEKYTDKLEKD---IAERNEELEAEKAKSEALLKM--------M 44
zc412hin     NL--MDHVFNMLETYASTLEEE---VSDRTKELVEEKKKSDVLLYR--------M 42
ruler        1.......10........20........30........40........50.....
```

Fig. 7. Sequence alignment of dimerization domains of guanylyl cyclases. Sequences encoding the putative dimerization domains of soluble and transmembrane guanylyl cyclases were aligned to identify the signature sequences of this domain of guanylyl cyclases

and concluded that the stoichiometry is 2 heme molecules per heterodimer. However, the same laboratory demonstrated that a bacterially expressed portion of the amino terminus of the β_1 subunit bound 1 molecule of heme and that the spectroscopic properties of this molecule closely resembled those of the enzyme purified from lung (Zhao and Marletta 1997).

Various approaches to determine the localization of the heme binding sites have been utilized. Deletion of the amino terminal 131 or 64 amino acids of either the α_1 or β_1 subunit, respectively, yields an NO-insensitive enzyme that retains basal guanylyl cyclase activity (Wedel et al. 1995, Foerster et al. 1996). Heme is no longer found associated with these molecules demonstrating that perturbation of the amino-terminal portion of either the α- or β-subunits disrupts heme binding and NO sensitivity. One model, therefore, is that the heme binding region is located in the deleted portion of the molecule. Alternatively, the structure of the molecule could be altered by the deletions such that heme is no longer able to bind. When a construct containing the amino-terminal 385 residues of the β_1 subunit was expressed in bacteria, it bound stoichiometric amounts of heme. Furthermore, as discussed above, the spectral properties of the heme containing deletion construct were nearly identical to those of the heterodimeric enzyme purified from lung, suggesting that the amino-terminal portion of the β_1 subunit contains the heme-binding domain (Zhao and Marletta 1997).

Histidine is the amino acid proposed as the ligand for the heme moiety. To identify the heme binding site, Wedel et al. (1994) mutated various His residues conserved in soluble guanylyl cyclases to Phe. Only mutation of His105 of the β_1 subunit abolished NO-stimulated production of cGMP, yet the mutation failed to eliminate basal activity. The His105Phe mutant no longer showed an absorbance maxima at 430 nm demonstrating that heme was not associated with the molecule (Wedel et al. 1994). Recently, mutations in the amino terminal deletion construct described above have been used to confirm the localization of heme-binding. Only mutation of His105 abolished heme binding to the bacterially-expressed deletion construct (Zhao et al. 1998). Thus, His105 of the β_1 subunit appears to represent the heme-binding site of soluble guanylyl cyclases.

The enzymatic activity of guanylyl cyclases is stimulated by a variety of factors including peptides and the gaseous molecules NO and CO (Drewett and Garbers 1994, Hobbs 1997). While peptides bind directly to their cognate transmembrane receptor guanylyl cyclase to stimulate enzymatic activity, NO and presumably CO bind to a prosthetic heme group bound noncovalently to the soluble enzyme (Ignarro et al. 1984, Kharitonov et al. 1995). One model for activation of the soluble enzyme, therefore, is that binding of NO to the Fe^{2+} of the penta-coordinate heme moiety leads to a conformational change that breaks the heme iron-His bond discussed above (Ignarro et al. 1984, Traylor and Sharma 1992). This conformational change results in as much as a 400-fold stimulation of guanylyl cyclase activity (Stone and Marletta 1996). Interestingly, protoporphyrin IX, which does not contain Fe^{2+}, stimulates soluble guanylyl cyclase independent of NO presumably by mimicking the nitrosyl-heme complex (Ignarro et al. 1982).

CO has also been proposed as a regulator of soluble guanylyl cyclase (Verma et al. 1993). The physiological significance of this signaling pathway is unclear, since CO stimulates the cyclase only minimally relative to stimulation with NO. It is postulated that the inability of CO to break the iron-His bond of the heme is responsible for its poor stimulatory properties (Kharitonov et al. 1995, Stone and Marletta 1995b). Interestingly, YC-1 has recently been shown to potentiate the stimulation of soluble guanylyl cyclase by NO, protoporphyrin IX, or CO, and elevates enzyme activity independent of these agents (Friebe et al. 1996). Remarkably, YC-1 renders the cyclase stimulable by CO to a level comparable to that of NO. These studies suggest that if an endogenous YC-1-like molecule exists, CO could in fact be an endogenous regulator of soluble guanylyl cyclase.

Whether CO is a normal endogenous messenger, however, remains unclear. The recent work showing that NO is cleared from hemoglobin as blood passes through the lungs (Gow and Stamler 1998) would support the concept that CO also would require a means to be rapidly and efficiently removed from hemoglobin. The conserved cysteine residues of hemoglobin that facilitate NO removal would not function in such a manner to remove CO. Thus, if CO is an endogenous signaling molecule, it clearly would be necessary that it function as such sparingly throughout the body.

Extracellular Ligand Binding Domain. Transmembrane guanylyl cyclases are divided approximately in half by a membrane spanning region. The putative ligand-binding domain of these molecules, therefore, resides outside of the cell (Fig. 1) and is the most divergent region of the transmembrane guanylyl cyclases. In the mammal seven membrane forms (GC-A-G) are known so far, and can be subdivided into several families according to their similarity within the extracellular domain (Wedel and Garbers 1998). For two of these families ligands have been identified. The natriuretic peptide receptors GC-A and GC-B bind ANP and BNP, or CNP, respectively. GC-C is the receptor for heat-stable enterotoxins produced by various pathogenic bacteria and possibly two endogenous peptides, guanylin and uroguanylin. For the family of neuronal guanylyl cyclases which are expressed in the retina (GC-E and GC-F) and in olfactory neuroepithelium (GC-D), ligands have not been discovered. Recently, a new membrane isoform has been cloned (GC-G); it is predominantly expressed in lung, intestine and skeletal muscle (Schulz et al. 1998). GC-G appears to be most closely related to the family of natriuretic peptide receptors based on sequence identity and the conservation of cysteine residues in the extracellular region. Several cysteine residues are conserved among receptor-like guanylyl cyclases as well as the natriuretic peptide clearance receptor. Cysteines in the clearance receptor that form intermolecular disulfide bonds have been identified (Stults et al. 1994, Itakura et al. 1994), and homologous cysteines can be identified in the extracellular domain of transmembrane guanylyl cyclases across many species, suggestive of conserved structure and possibly conserved function. Four of these cysteines are nearly invariant in 26 extracellular domain sequences (Fig. 8).

Extracellular domains are also modified heterogeneously by glycosylation. For example, natriuretic peptide receptors are variably glycosylated, GC-E contains only one predicted N-glycosylation site, and GC-F is not predicted to be N-glycosylated (Yang et al. 1995). The role of glycosylation is not clear, however, there is a positive correlation between the glycosylation state of natriuretic peptide receptors and their responsiveness to ligand (Koller et al. 1993, Fenrick et al. 1996). Although interpretation of these results must be made with caution given the nature of the experiments, the data are consistent with glycosylation being important for ligand binding, either because of their requirement for appropriate protein folding or for their effects on protein conformation after folding.

Although sequences encoding extracellular domains of transmembrane guanylyl cyclases are the most divergent, there are stretches of amino acids retained across the various extracellular domains. Since ligands for only three mammalian cyclases have been identified, a major question in the field is whether or not the remaining cyclases actually possess ligands. To examine whether an orphan cyclase could be regulated by a ligand, Baude et al. (1997) constructed a chimeric receptor between the extracellular and kinase homology domains of the CNP receptor (GC-B) and the dimerization and cyclase homology domains of a *C. elegans* orphan receptor, GCY-X1. CNP stimulated the guanylyl cyclase activity of the chimeric receptor with a dose response curve similar to that of wild-type GC-B (Baude et al. 1997). These results suggest that orphan cyclases can be regulated by extracellular ligands, and support the hypothesis that these cyclases are indeed receptors for unidentified ligands.

```
C52E4EXT  ------------------------SRL-------------SSSDLAV--SRVETSKDSDGETPRPTSSELKEVNRIREEALAQEKEEERTTKEENQKIEE 61
DRGOEXT   VTD--------------------G----------GLP-----------LAIEDVNKNPNLLPGKK----------------------------------- 24
DRG1EXT   MTRWP----------------------FNLLLLLSVA---VRDC-----------------------------------SNHRTVLTVGYLTA---LTGD 37
EGCBEXT   MDLGHSL--------FVVFTC-----------------FLMARCRTEIGKNIT---------------------VVVMLPDNHLK-Y---SFAFPRVFPA 50
FGCCEXT   MRKEH----------------TW----LCLQIILWLTDLL------EA--------NCMSGSLTMNVIMLNDSMTEWNIKAVQEAVSIG----------- 55
GCX1EXT   M---------LR--WLTLLSC-----------------ILLTALHGNIVEDVGAAQQASYPPPPPPF----PFNVIVILPKREST-YDNFGMTLQKAMPV 67
GCY1EXT   M-------------QIFTI-------LL-LFN-IFPSIFVQNLPDTTVAPRTKRTIKIGIAAAQ-----------------RIQTSSIGW---------- 51
GCY10EXT  ML----------------------KSLLIIVIVFLHRELCDGIQLILFDNWPSAQNVCASAVADATANGQCTTKSIHSFSTTAKNVKNRVTRKHLELLTV 78
GCY12EXT  M--WPTSVEALRFYHTFAFSPGRRRLFGLSIVFVIAALCVTSCDADAVPAPSPEAEPSLGAGANPYLLDSLVKRLPKKGDKKEIL-ISYLAAVPTLMGE 97
GCY2EXT   MV-----------SSILKF-------VI-LIHSTFHSTFAQNLPDTTVAPKTKRTIKIGIAAAQ-----------------RTQTSSIGW---------- 54
GCY3EXT   M------------KNVFQL-------LIPLFFHLFSLVSLQNIPVSTGTTRPK--LKVGIAAAQ-----------------KTQSASIGW---------- 52
GCY4EXT   MIL---------FSKYYFS-------MRQLNYYIFISTILTY--NLTHGQGPRPVIRVGITAAL-----------------KTENGSIGW---------- 55
GCY5EXT   MRL---------L---YFS-------MVLL----WVLGASEC--QVIPS--SRRTLRVGIAAAQ-----------------DTQSGSIGW---------- 46
GCY6EXT   MIG---------VYLRSVI-------FPL--------LFVQTICQPPGNVFHLGFLHCDVLENNV----------------EGSTTYINY---------- 50
GCY7EXT   MKP---------FYSMSLV-------LFLVITLLPKPMFPQVATGTTGNVIRVGFIHC----RDF----------------QSAPITVGY---------- 54
OLG3EXT   MMH-------------------------------------------------------------FCFLLVCPCL----LSITSVH--------------- 20
OLG4EXT   MQH-ISPNGWES--NHPCVS------------------------------------------------IERNRRTLQSLPFYNIMLWL-LL--G--VLTF 44
OLG5EXT   MTN-HRDTCRLH--NLAYHPPWW---------------HSTTKMKRNKDTRTTSSSEGCRQLKTLSALLLTSSLSKPQLPFSSMHLFRS--WKE--KLSL 78
RGCAEXT   MPGSRRVRPRLR--ALLLLPP-----------------LLL--LRGGHASDLT---------------------VAVVLPLTNTS-Y---PWSWARVGPA 54
RGCBEXT   MALPSLL--------LVVAA-------------------LAGGVRPPGARNLT---------------------LAVVLPEHNLS-Y---AWAWPRVGPA 48
RGCCEXT   MTSLL----------------G-----LAVRLLLFQPTLMFWASQVRQ--------KCHNGTYEISVLMMDNSAYKEPLQNLRDAVEEG----------- 60
RGCDEXT   MAG-LQQGCHPE--GQDWTAPHW---------------KT------------------CRALPGPRGLTVRHLRTVSSISVFSVVFWGVLLWAD--SLSL 62
RGCEEXT   MSAWLLPAGGFP--GAGFCIPAW---------------QSRSSLSR------------------------------------------VLRWPGPGLPGL 41
RGCFEXT   M-F-LGPWPFSR--LLSWFA------------------------------------------------ISSRLSGQHGLTSSKFLRYLCLL--A--LLPL 44
RGCGEXT   MASRARSEPPLE--HRFYGGA-----------------ESHAGHSSLVLTLFVVMLMTCLEAAKLTV----GFHAPWNISH---------PFSVQRLGAG 68
SPGCEXT   MAH--------------------ARHLFLFMVAFTIT-----------MVIARLDFNPTIINEDRG----------------RTKIHVGLLAEWTTADGD 53
ruler     1.......10........20........30........40........50........60........70........80........90.......100
```

Fig. 8

```
C52E4EXT  V-------------------------------------------GEDHVSEATSLLDSEVSHGDNNISFSQMPSDSIPHEDRTS---------------- 102
DRGOEXT   -------------------------------------LAFKPV--DI-----------GHKMSAYRVK------------PLRAMTQMREA-GVTAFIGP 61
DRG1EXT   L-----------------KTRQGLA------------------ISGALTMALDEVNKDPNLLPNVYLDLRWNDTKGDTVLATKAITEMICD-GIATIFGP 101
EGCBEXT   IRMAHDDIQKKGKLLR------GYTINLLN--HS-TESQGAGCSESQAQIMAVDTKLYEKPD-------AFFGPGCVYSVASVGRFVNHWK--LP-LITA 131
FGCCEXT   MHVVTKDLEREGIKVTINADFQTFNTDLYAT------------PG-CVSSGCEGVEKLKNLRHTRRLGCVILGPTCTYAT--YQMLSLKNTFGVPLISAG 140
GCX1EXT   IDIAVQEVIKAKKLPP------GW-INLTY-WDS-RLYEDILLAERHATVGVIQAYCEHRLD-------AILGFADNYGLATVTKVTAGLNGGIP-ILTT 150
GCY1EXT   -----------------------------------------SVCGGAVPLAIERL-KEMGFVKDFDFEYIVDYTECDLGSVVRAGMEFIKTHKVDVIIGP 109
GCY10EXT  IIL----------------------LKLFGVFHRINQQH--GCSGDNSVKSASYAINAVASRTSGELDFVFVGPTCTTDIRTIGDF-------------- 140
GCY12EXT  IDQYLLNLSASNSAQNSSKDRESFSKKLNQIVHCLTTDAYVSVVSGALVAAIVEINQDKDLIPAYQLKYVFGNTCGNDSHSTRLFMEHWQA-GARVFIGP 196
GCY2EXT   -----------------------------------------SVCGGAVPMAIERL-REFGYVKDFDFEFIVDYTECDQGSVVRAGIEFIKTHKVDVIIGP 112
GCY3EXT   -----------------------------------------NVCGGAVPLAIERL-KQAGYVTNFDFEYYVEYTECDLASTVRTGINFLKNLEVDVIIGP 110
GCY4EXT   -----------------------------------------AYTGGAVPLALQYL-KSHGYMLNFDFEFHVEYTECDLANTVRAGLNFMKTNNYDVIIGP 113
GCY5EXT   -----------------------------------------ASCGGTLPIAVQYL-KSKGFLTDFDVEYYMEYTECDRASVAKAGMKFMKEMNVDVVVGP 104
GCY6EXT   -----------------------------------------RTSASAASIAIDKI-KREGLLLGYDFKFTILYDQCDENIAAGNAIKLFAEHNVDVLFGP 108
GCY7EXT   -----------------------------------------RTSAAAASIAVDRL-KRENLMSGWEFNFTIEFDDCVESEAAGMTVDLIEKHNVDVIIGP 112
OLG3EXT   --------------AA------TFKLALVGPWSC-DPMFSRAMPSAAANLALSRLRSDADLSRGFWYDVKLLNEDCS-ASKALTELGEMEGYGHA-YIGP 97
OLG4EXT   ---------PCCVHCL------IFKVGVLGPWNC-DPVYYRALPTVAARLAISRINRDPNLDLGLTMDFIILQEPCE-TSKAHTTFIYYDKSANA-PVGP 126
OLG5EXT   LFLFIISFFPCQIWAT------TFKVALVGPWTC-DLLYSKALPDLAARLAISRINKNPYLNRGYWYDYALVNEDCK-STRALVRFSDLEGYGAA-FLGP 169
RGCAEXT   VELALARVKARPDLLP------GWTVRMVL--GS-SENAAGVCSDTAAPLAAVDLKWEHSPA-------VFLGPGCVYSAAPVGRFTAHWR--VP-LLTA 135
RGCBEXT   VALAVEA------LGR------ALPVDLRF--VS-SELDGA-CSEYLAPLRAVDLKLYHDPD-------LLLGPGCVYPAASVARFASHWH--LP-LLTA 122
RGCCEXT   LDIVRKRLREAELNVTVNATF-IYSDGLIHK------------SGDCRSSTCEGLDLLREITRDRKMGCVLMGPSCTYST--FQMY-LDTELNYPMISAG 144
RGCDEXT   ---------PAWARE-------TFTLGVLGPWDC-DPIFAQALPSMATQLAVDRVNQDASLLLGSQLDFKILPTGCD-TPHALATFVAHRNTVAA-FIGP 143
RGCEEXT   LLLLLLP--SPSAFSA------VFKVGVLGPWAC-DPIFARARPDLAARLATDRLNRDLALDGGPWFEVTLLPEPCL-TPGSLGAVSSALTRVSG-LVGP 130
RGCFEXT   ---------IWWGQAL------PYKIGVIGPWTC-DPFFSKALPEVAAALAIERISRDMSFDRSYSFEYVILNEDCQ-TSKALTSFISHQQMASG-FVGP 126
RGCGEXT   LQIAVDKLNSEP-VGP------G---NLS--WEF-T-YTNATCNAKESLAAFIDQVQREHIS-------VLIGPACPEAAEVIGLLASEWD--IP-LFDF 144
SPGCEXT   QG----------------------------------TLGFPAL--GALPLAISLANQDSNILNGFDVQFEWVDTHCDINIGMHAVSDWWKR-GFVGVIGP 116
ruler     .......110.......120.......130.......140.......150.......160.......170.......180.......190.......200
```

Fig. 8

```
C52E4EXT  ---------------------------------------LPSATPSEIGDAISKKKL-----------------------------------------EK 122
DRGOEXT   --DESCTTEALLASAWNTPMLSFKCSDPIVSNKS-TFHTFARTLAPASKVSKSVISLLNAFHWNKFSIVV------SSK-----PIWGSDVARAIQELAE 147
DRG1EXT   --EGPCYVEAIVSQSRNIPMISYKCAEYRAS----AIPTFARTEPPDTQVVKSLLALLRYYAWNKFSILY------EDVWSPVADLLKDQATKRNMTINH 189
EGCBEXT   WAPAF-GFDSK-EEY--RTIVRTGLSTTKLGEFAHYL------------H--------SHFNWTTRAFMLFHDLKVDDR---------------PYYFIS 192
FGCCEXT   SFGLSCDYHRSLA----------------------------RMLLPARKITY----FFKEF-WQ------YEDF-IKPKKWQSVYIYKWDGNTESCFWYI 200
GCX1EXT   SGMPS-LLNSKKE-Y--PFLTRMQGSYRLLADSMYQLIAYHDEDSVSKSN--------SSLNYLNLIFFYHDKRRAVNRVIAQDESQETGATSSHCYFSL 238
GCY1EXT   PCAQALRLMSFLAENYKKPVLGWGFVSDTDLSDVIRFPHLTTVIPNSLMLGYAASKMLTTFHWNRVALLY----------------YFS-DVKYCSGVMN 192
GCY10EXT  ------------AEIWKSPVIGYE-------------PVFEARGVQELTSVINVAQF-----------------------------------------SV 174
GCY12EXT  --EKNCKTEAAMAASQNLPIISYRCNDQDISRDDYHYRTFARTVPPAGEIFKAFMSLMKQYNWRKFSVVY------DVKKGQVKNELFE-TLKRMVETEN 287
GCY2EXT   PCAQALRVMSFLAENYKKPVLGWGFVSDTDLSDVIRFPYLTTVIPNSLMLGYAASKMLTVYNWGRVAMLY----------------YYS-DIKYCSGVMN 195
GCY3EXT   PCSEAIRTMATLATLYKKPVLGWGFVSQADLSDMTRFPYLITVLPTSQTLGYAASKLLELYKWDKVALLY----------------YKS-EVNHCGGVMN 193
GCY4EXT   PCAPALKMMGTLSTIYKKPVLGWGFVSESEISDMNRFPFVASVLPSTKTLGVVTSKLLEIYGWDRVALLY----------------FKN-ELDYCSSVVN 196
GCY5EXT   SCGDALAIMGTLSAIYKKLVLGWGFVSDTQLADTNRFPYVASVQPTAQTLGLATSRILEMFQFDRVALLY----------------YKD-DQDYCKSVMD 187
GCY6EXT   TTNNAAMPVFILATYYNIPLITWGITSSATLDDESRFPTAGMLSIGSRSLAVTFREVMKEYGWDQFVFAY----------------SLEMNDEKCETLRD 192
GCY7EXT   TMNQPTLAAFIVSNYFNRPIIAWGLVNAAQLDDFERFPNAGILSAGQRSLGVAIRAVLKRYEWSQFVYAY----------------FTEEDTEKCVTMRN 196
OLG3EXT   FNPALCHAASLMAQHWDAGLASPGCLDPVWLN--------LPPITPPSKV---LFTVLRFFRWAHVAVA-----------SAPTDLWESTAQEVASCLRA 175
OLG4EXT   TNPGYCIAASLLAKSWDKALFSFSCISYE-LDRLTAYPTFARAAPFPADV---LFTVFKHYRWATSVVI-----------SSNEEIWIETAGRVASALRM 211
OLG5EXT   ANPGYCSSAALYTKEWDVGILSWGCLKP-YMKTVDMFPTLLRPLPLSSNV---LFTVLRYFRWAHVAII-----------SEDTDLWEATGHELASSLRA 254
RGCAEXT   GAPAL-GIGVKDE-Y--ALTTRTGPSHVKLGDFVTAL------------H--------RRLGWEHQALVLYADRLGDDR---------------PCFFIV 196
RGCBEXT   GAVAS-GFAAKNEHY--RTLVRTGPSAPKLGEFVVTL------------H--------GHFNWTARAALLYLDARTDDR---------------PHYFTI 184
RGCCEXT   SFGLSCDYKETLT----------------------------RILPPARKLMY----FLVDF-WK------VNNAPFKTFSWNSSYVYKNGSEPEDCFWYL 205
RGCDEXT   VNPGYCPAAALLAQGWGKSLFSWACGAPE-GG-----GALVPTLPSMADV---LLSVMRHFGWARLAIV-----------SSHQDIWVTTAQQLATAFRA 223
RGCEEXT   VNPAACRPAELLAQEAGVALVPWGCPGT------RAAGTTAPAVTPAADA---LYVLLKAFRWARVALI-----------TAPQDLWVEAGRALSTALRA 210
RGCFEXT   ANPGYCEAASLLGNSWDKGIFSWACVNHE-LDNKHSYPTFSRTLPSPIRV---LVTVMKYFQWAHAGVI-----------SSDEDIWVHTANQVSSALRS 211
RGCGEXT   VGQMT-AL---EDHFWCDTCVTLVPPKQEIGTVLRESLQY--------------------LGWEYIG--VFGGSSAGSSWGEVNELWKAVEDELQLHFTI 218
SPGCEXT   --GCGCTYEGRLASALNIPMIDYVCDENPVSDKS-IYPTFLRTIPPSIQVVEAIILTLQRYELDQVSVVV------ENI-----TKY-RNIFNTMKDKFD 201
ruler     .......210.......220.......230.......240.......250.......260.......270.......280.......290.......300
```

Fig. 8

```
C52E4EXT   EDSNSS----MSSLDERTTVSAKPTTTRRLLNQKDLEKEK-KRSSMAGSSVVILC--------------------------------------------- 172
DRGOEXT    A---RNFTISH--FKYIS----DYIPTTKTLS---Q-IDKIIEETYATTRIYVF---IGEHIAMVDFVRGLQNRRLLESGDYIVVS--VDDEIYDSN--R 227
DRG1EXT    K---QSFIDNR--VKCCE-QMLDCCRSGYWYQL--------VQNTMNRTRIYVF---LGAANSLVDFMSSMETAGLFARGEYMV--------IFVD---- 260
EGCBEXT    EGV---FL----VLRRE-NITVEAVPYDDQ---KNSDYREMISSLKSNGRIVYICGPL-----DTFLEFMRIFQNEGL-PPEDYAIFYLDMFAKSILD-- 273
FGCCEXT    NALESGVSYFNNALKFKEILR-------------TEGELMKVLQENNHKSNVILMCGTPND---------IWNLHNKVAIPQDKVLILLDI----------- 269
GCX1EXT    YAIKRYFTEKSKTFKREWALNTPQFPFDEDLVIERETFKQWLREISLQSNVIILCASP-----DTVREIMLAAHDLGMATSGEYVFINIDV-------ST 326
GCY1EXT    D------IEAT--FNDPSTPNVNIVIKAEIYLNDNETTDNVFQTVKSRARIILWCTQTSVEK---RDYLIKIATHDMIGDEYVHIMLSMRNVAFGTQTSL 281
GCY10EXT   GGVAETLVFLMKELEQVEIMNSFKIREYVEVDENDVDWTKVDQKIKRGARMIVVCAD--------------------------------------------- 231
GCY12EXT   K-----FEEHK--FEIMNVSKLEFSKMDISSQQDIQSIENAIKSTMQTTRIYLT---FDNVRLFRTMLSIMGEMGL-TQQDYML---------IYVDTNYD 368
GCY2EXT    D------VEAT--FNNPSTPNVNIVIKAEIYLNDNETTDIVFQSVKSRARIIFWCTQTAIEK---RDYLIKIATHDMIGDEYVHIMLSMRNIAFGAQTSL 284
GCY3EXT    D------VETT--FNDPSTYSIQIVLKAEIDASDNVTTNAVISSVKTRARIILWCAQLGSEK---RDYMIKISQLGLDTDEYVHVLISMRSIGFGVQTFV 282
GCY4EXT    D------VEKS--LYNESK-SVQIVMKAEIDGSNSESTSATLQVVKTRARVILFCAHAGGEK---RFYLIQAGLLGMNSSEYVHVMLSMRSVGFGVQTSV 284
GCY5EXT    D------VEAT--LSDPDLYPVRIVWKGELQSDNEALTRSTLQAVKSRARIVLLCAISGPEK---RNYLISIAQQNMTTNEYVHILLTMRSIGYGVQTSL 276
GCY6EXT    D------FQNM--VAYYG--DIVLSYAVQIMDHSEEGLLAILKDVSTRGRIIVPCFHEGNSRGLHRRWMLVAARNGFVNDEYVYIFPSLRSRGYAV---- 278
GCY7EXT    D------LQQV--VSYFG--DIILAYSIQVADISNDGMIEALKKIQSRGRIIVTCMKDGI--GLRRKWLLAAEEAGMIGDEYVYVFSDIKSKGYVV---- 280
OLG3EXT    MGL---------------PVGPVVSMETNSKDGAQEALYQVKTA--DKVNVIIMCMSSVLIGGEHQRELLLTALDMGM-ISDGYVFIPYDALLYAM---- 253
OLG4EXT    KGL---------------PVGFVAAMGMNDTE-LESTLKKIQST--GGVRVIIMCIDSVLVGGEQQAVFLLKAKQLGL-TSGKYVFVPYDTLHYSV---- 288
OLG5EXT    LGL---------------PVNPVVTMETE-KDGPRSALAKVREA--DRVRVIIMCMPSVLIGGQAQYQLLTTALAMRM-IDRGYVFIPYDTLLYSL---- 331
RGCAEXT    EGL---YM----RVRERLNITVNHQEFVEG---DPDHYPKLLRAVRRKGRVIYICSSP-----DAFRNLMLLALNAGL-TGEDYVFFHLDVFGQSLKSAQ 280
RGCBEXT    EGV---FE----ALQGS-NLSVQHQVYTRE---PGGP-EQATHFIRANGRIVYICGPL-----EMLHEILLQAQRENL-TNGDYVFFYLDVFGESLRAGP 266
RGCCEXT    NALEAGVSYFSEVLSFKDVLR-------------RSEQFQEILMGRNRKSNVIVMCGTPET---------FYNVKGDLKVADDTVVILVDL----------- 274
RGCDEXT    HGL---------------PIGLITSLGPGEKG-ATEVCKQLHSV--HGLKIVVLCMHSALLGGLEQTVLLRCAREEGL-TDGRLVFLPYDTLLFAL---- 300
RGCEEXT    RGL---------------PVALVTSMVPSDLSGAREALRRIRDG--PRVRVVIMVMHSVLLGGEEQRYLLEAAEELGL-TDGSLVFLPFDTLHYAL---- 288
RGCFEXT    HGL---------------PVGVVLTSGQDSRS-IQKALQQIRQA--DRIRIIIMCMHSALIGGETQTHFLELAHDLKM-TDGTYVFVPYDVLLYSL---- 288
RGCGEXT    TARVRYSSGHSDLLQEG------------------------LRSMSSVARVIILICSS-----EDAKHILQAAEDLGLN-SGEFVFLLLQQLEDSF---- 284
SPGCEXT    E---RDYEILH--EEYYA----GFDPWDYEMD---DPFSEIIQRTKETTRIYVF---FGDASDLRQFAMTALDEGILDSGDYVILGAVVDLEVRDSQDYH 286
ruler      .......310.......320.......330.......340.......350.......360.......370.......380.......390.......400
```

Fig. 8

```
C52E4EXT   ----------------------------------------------------------------ISKCIYS---KSSFQTSS--SAHSHSIRSKKDT--- 200
DRGOEXT    RVN--I-ME-RNYLDP-------------YIRKE-KSKSLDKISFRSVIKIS-------------MT----YPQNPHIRVPIYG------------LH-- 278
DRG1EXT    -MMVYSEREAEKYLRRVD---QITFMSNCHSTENFNQM-----ARSLLVVASTPP---TKDYIQFTKQVQKYSSKPPFNLEIPRLFVESNFSKFISIYAA 348
EGCBEXT    --KD-----------------YKPWESS-DINWTD-PIKL---FKSVFVITAKE-PDNPEYKAFQRELHARAKQEFSVQ------LEPS----LEDIIAG 338
FGCCEXT    ----------------FNTVYY------------DNKS-SPYYMENVLVVTQRPSNMSKISNQ------------------------TGIAKLLEDNYAA 316
GCX1EXT    GSHA-----------------EQPWIRANDTNNEE-NEKPKEAYRALKTISLRR-SDLDEYKNFELRVKERADQKYNYT------NITGKDYEMNNFISA 401
GCY1EXT    GKPTFSQSGLTPIWESFT------------EGTDGFEKMAKQAATRMFVLDVNSEVADKKYLEYMQKNIIKAVQSPPMNCSTVECMTANTTI--MGGYAR 367
GCY10EXT   ------------FYDIYSAFYNIGIRSLS-------------GFRFIIVVILNKPPDEILNQPNVKNLLYG---SNAFIISPLQEQYSDAFSIMQDVIPN 303
GCY12EXT   WLNVYHSMN-NHFLR--N---TMTYLHHSWDANNSSDRKMLDYARSALSIIPTPVKLNSQRFYNFWKKAGDY--MHHFGVQKADNLKGNRI-------AC 453
GCY2EXT    GKPTFSQSGLTPIWESFT------------EGTDDFEKMVKQAATRMFVLDVNSEVADKKYLDYLQKNIMKAVQSPPMNCSTVECMTANTTI--MGGYAR 370
GCY3EXT    GKTTFKLSGLTPVWESFS------------NVTDGLENVAKRGATNVLVIDLNSEVSDKAYLDYMQRNILNVVKLPPLNCSTVDCVSSTATG--MGAYAR 368
GCY4EXT    GKKPLTLSGLPPIWESFT------------VNPDGMEDLAKSVAAKMIVIDTSSEVRDKTFLQYMTKNIVYAIREPPLSCKVPECLITNATG--MGAYAR 370
GCY5EXT    GKKTFA-NGLTPLWESFT------------VAPDGNETNARRAAEKMLVIDVNSDVQDAEFLQYLTKNIADAVRNPPMKCNTSECINASSTS--MGSYAR 361
GCY6EXT    ---PQADGTFRYPWTEAT------------GPQPGDQEALLGFQKSIFIVDMQGQGNVGSNYTQFEHEIIQRMKEPPYNC-TDACASPEYQT--AATYAG 360
GCY7EXT    ---PLLGGGERPSWILST------------GSDENDTRALKAFKQSIFICDMMGQGSIATNYTIFGQEIIARMKEAPYFC-TKDCEGENFTV--AATYAG 362
OLG3EXT    -----------------------PYQDMTFHQLTN-STQLRHAYSAVLTVTMES-DQ-NFYQAFRQ-----AQMSREIR------SAISAT-EVSPVFGT 315
OLG4EXT    -----------------------PYTNVSHFALKN-NSSLKEAYDAVLTITMASEPL-SFNEAFEA-----ARKSGEIT------LPVQPE-QVNPLFGT 351
OLG5EXT    -----------------------PYKDAFYYMLGN-DTKLRRAFDGVLTITMESGER-NFYEAFKD-----AQDRFELR------SSTPTE-QVSPFFGT 394
RGCAEXT    GLVP-----------------QKPWERG-D---GQ-DRSARQAFQAAKIITYKE-PDNPEYLEFLKQLKLLADKKFNFT------VEDG----LKNIIPA 347
RGCBEXT    TRAT-----------------GRPWQDN-RTQ-EQ-AQALREAFQTVLVITYRE-PPNPEYQEFQNRLLIRAREDFGVE------LAPS----LMNLIAG 335
RGCCEXT    ----------------FSNHYF------------EDDTRAPEYMDNVLVLTLPPEKFIANASV------------------------SGRFPSERSDFSL 322
RGCDEXT    -----------------------PYRNRSYLVLDD-DGPLQEAYDAVLTISLDTSPE-S--HAFTA-----TKMRGGAA------ANLGPE-QVSPLFGT 361
RGCEEXT    -----------------------SPGPEALAAFVN-SSKLRRAHDAVLTLTRRCPPGGSVQDSLRR-----AQEHQELP------LDLDLK-QVSPLFGT 352
RGCFEXT    -----------------------PYKHSPYQVLRN-NQKLREAYDAVLTITVESHEK-TFYEAFTE-----AAAGGEIP------EKLDSH-QVSPLFGT 351
RGCGEXT    ------------------------WKEVLAED--K-VTRFPKVYESVFLIAPSTYGGSAGDDDFRKQVYQRLRRP-PFQ------SSISSEDQVSPY-SA 349
SPGCEXT    SLD-YI-LDTSEYLNQINPDYARLFKNREYTRSD-NDRALE--ALKSVIIVTGAPVLKTRNWDRFST----FVIDNALDAPFNGELELRAEIDFASVY-- 375
ruler      .......410.......420.......430.......440.......450.......460.......470.......480.......490.......500
```

Fig. 8

```
C52E4EXT   ----------RDKSRCK--C------EDIRADNKLKTKVCSIMNTD--LDTL----------GISLDDFS------KALPYHF----------V-----I 249
DRGOEXT    -LYDSVMIYVRAITEVLRLGGDIY----------DGNLVMSHIFNRSYHS-IQGFDVYIDSNGDAEGNYTVITLQNDVGSGASIGS-----LAKMSMQPV 361
DRG1EXT    YLYDSVKLYAWAVDKMLREETRVLTDDVIFEVSSNGTRVIDTIIKNRTYMSITGSKIKIDQYGDSEGNFSVLAYKPHKWNNSNNMPCNYHMVPV----AY 444
EGCBEXT    CFYDGFMLYAQALNETLAEGGSQNDGINITQKMQNR--R---------FWGVTGLVSTDKNNDR-DIDFNLWAMTNHKTG-QYGIVAYYNGTN------- 418
FGCCEXT    GYLDGVLLFGHILKKFLGSVDI------------NQTFSFIDQFRNISIIGALGPLILDAAGDRELNLTLLYS---------STATNNYTEL-------I 388
GCX1EXT    -FYDAVLLYAIALNETIQSGLDPRNGHNITSRMWGR--T---------FVGITGNVSIDHNGDR-YSDYSLLDLDPVQN--RFVEVAYYSGAS------- 479
GCY1EXT    HLFDVVYLYGIALTHTNSTDP------AVYG---DVDVLVHQFVTS--FQGMTG-HVVISPNLTRMPIFQLYGLNSD-----------YDQVALVDF--- 441
GCY10EXT   LADDQFTTFLRIYHACYAYCVGSVNGAETQTDNYHTAMSGKAVTTK--YGTFT-----FDNSGSVLTNYAVFTINPAEMTFES----------ILTLKSV 386
GCY12EXT   YLYDAVYLYAKAIHELVEE----YGNDDSYDPTADGKAIIDRIVNKK-YRSIQGFDMRIDERGNSKGNFSLLSWQKVAPIMNKSDPSYYPLDHALDLTAI 548
GCY2EXT    QLFDVVYLYGVALTNTNSTDP------AVYD---DVDVIVPQFVTS--FQGIE--------------------LCNS-----------F---------- 417
GCY3EXT    HLFDVVYLYGIALTRVNSTDS------AVYD---DMSKLIPQFVTS--FNGMTG-LVAINQNLSRMPLYQLYGLDEQ-----------YEQTSLMNL--- 442
GCY4EXT    HLFDVFYMYGMAVSSLNSTDP------NVYG---NLSLLIPKFTTA--FEGMTG-EVKLNEDLARKPLYQVYGLNSE-----------YDQISLIDI--- 444
GCY5EXT    HLFDVFYLYGMAVSKLNSTDP------TVYG---NLNLLMPQMVTS--FDGMTG-RVQIGQNLYRVPTYQLYGLDEK-----------YEQVALVNM--- 435
GCY6EXT    QLHDSVYIYGVVMDQIMKTVP------NQYK---NGTAFPRKMAGV--FNGVGG-TVAIDEGGGLQPTLFVLTLDSN-----------NNSSLIMTV--- 434
GCY7EXT    QLHDAVYAYGVALDKMLKAGQI-----AQYR---NATAFMRYFPQS--FIGMSG-NVTINEKGTRNPTLFLLALDEN-----------NNNTRMATI--- 437
OLG3EXT    IFNMVYFVAKAVEERRLAGGGHWVTGSQLFQSD--GGFE---------FQGFNQVLYGGKGGRGLQARYVV--LD--SEGDRLVPTHSLAPTH------- 393
OLG4EXT    IYNSIYLMARSIHNARKAGMT--LSGSNVAYYT--KNTN---------FDGFNQRFEVDSRGQVKTS-YVV--LDSDSKGAELYQAYVV----------- 424
OLG5EXT    IYNMMYYIAMAVEQTR-ASGVHWVTGRSLGNSE--GGFE---------FEGFNQPLQAGRNGEGMQAGYVV--LDYSGMGNSLYSTHLLHPTH------- 473
RGCAEXT    SFHDGLLLYVQAVTETLAQGGTVTDGENITQRMWNR--S---------FQGVTGYLKIDRNGDR-DTDFSLWDM-DPETG-AFRVVLNYNGTS------- 426
RGCBEXT    CFYDGILLYAQVLNETIQEGGTREDGLRIVEKMQGR--R---------YHGVTGLVVMDKNNDR-ETDFVLWAMGDLESG-DFQPAAHYSGAE------- 415
RGCCEXT    AYLEGTLLFGHMLQTFLENGES------------VTTPKFARAFRNLTFQGLEGPVTLDDSGDIDNIMCLLYV---------SLDTRKYKVL-------M 394
RGCDEXT    IYDAVILLAHALNHSETHGTG--LSGAHLGNHI--RALD---------VAGFSQRIRIDGKGRRLPQ-YVI--LDTNGEGSQLVPTHIL----------- 434
RGCEEXT    IYDAVFLLAGGVTRARAAVGGGWVSGASVARQM--REAQ---------VFGFCGIL-----GRTEEPSFVL--LDTDAAGERLFTTHLLDPV-------- 426
RGCFEXT    IYNSIYFIAQAMSNALKENGQ--ASAASLTRHS--RNMQ---------FYGFNQLIRTDSNGNGISE-YVI--LDTNGKEWELRGTYTV----------- 424
RGCGEXT    YLHDALLLYAQTVEEMMKAEKDFRDGRQLISTLRADQVT---------LQGITGPVLLDAQGKR-HMDYSVYALQKSGNGSRFLPFLHYDSFQ------- 432
SPGCEXT    -MFDATMQLLEALDRTHAAGGDIY----------DGEEVVSTLLNSTYRSKTDTFYQF-DENGDGVKPYVLLHLIP-IPKGDGGAT-----KDSLGMYPI 457
ruler      .......510.......520.......530.......540.......550.......560.......570.......580.......590.......600
```

Fig. 8

```
C52E4EXT   D--------ESC------------------------------------------------------------ 253
DRG0EXT    GFFAYDKNSVIPEFRYIKND-----RPIQWL--N-GRPP--LAEPLCGFHGELCPRKKLDWR-YLVSGPLCA 422
DRG1EXT    -FHQGEEH---PEYKLING-------SIDWP--SGGEKP--ADEPMCGFANELCKKDDTH---Y--TSTVAA 496
EGCBEXT    ------KEIV----WS-------ETEKIQWP--KGSP-P--LDNPPCVFSMDEPFCNEDQ---LPVLG---- 461
FGCCEXT    QFDTSTNQTTVMDTSPNF-------------IWKNHRLPS--DVPQSG-------------PH--------- 423
GCX1EXT    ------NQLK-------------TVGQLHWV--GGKP-P--TDLPICGYDKSK--CPGYPLHVYLLMG---S 522
GCY1EXT    ---TYTNSVMVPNVTLFYKDEGGAV----WSYY-GHSRP--LDIPICGFLGKSCPVS--FWEQY-------K 494
GCY10EXT   A--------KSCDTYNCFQLSPNKTSDLLWTLKDMDPPDDCVAKSSC-------------------VNYIPH 431
GCY12EXT   -FVEAPDKDRLPNLQFKSS-------RIQWL--KG-EPP--PDEPVCGFHGENCRKKG-------------- 593
GCY2EXT    ----------VDNVTLSYKEEGGAV----WYFY-GNSRP--LDIPICGFLGKFCPIS--FWEQ--------- 461
GCY3EXT    ---SFADGTTVATISLAYSNESSAV----WHFW-GGIRP--LATPICGFLGNSCPLP--FWEQY-------G 495
GCY4EXT    ---SLINDTV--NVVFNYADGNTEV----WHFW-GGTRP--PDTPVCGFLGKSCPLP--IFEQY-------- 494
GCY5EXT    ---TFYNSSS--QLSRGYSDEGRSV----WHFW-DGTRP--LDTPICGFSGRYCPVQ--FWDQY-------G 486
GCY6EXT    ---DVDQQEAV--VTKHYTNEATAL----WSHRKG-IRP--PDQPICGYTGSLCPAN--------------- 479
GCY7EXT    ---YVENMSAT--FNALYSDEGV-M----WASRKNNARP--VDVPLCGFTGNLCPKS--FVDEY-------- 487
OLG3EXT    ------TDGTVGGLRP-------LGRSFFFP--GGKP----SKSSFCWFSPEE-ACSGGLDAVS-------- 437
OLG4EXT    -------DLRSGVLRF-------TGRSIHFP--GGSP-P--VADSSCWFEENA-VCTGGVEVTY-------- 468
OLG5EXT    ------TDTMTGGLRY-------LGRSVHFA--GSTP----YTDSSCWFSPYF-ACSGGWDSTTA------- 518
RGCAEXT    ------QELM----AV-------SEHKLYWP--LGYP-P--PDVPKCGFDNEDPACNQDHFSTLEVLA---- 472
RGCBEXT    ------KQI-----WW-------TGRPIPWV--KGAP-P--LDNPPCAFDLDDPSCDKTP---LSTLA---- 457
RGCCEXT    AYDTHKNQTIPVATSPNF-------------IWKNHRLPN--DVPGLG-------------PQILMIA---- 434
RGCDEXT    -------DVSTQQVQP-------LGTAVHFP--GGSP-P--AHDASCWFDPNT-LCIRGVQP-------L-G 478
RGCEEXT    ----------LGSLRS-------AGTPVHFP--RGAPAP--GPDPSCWFDPDV-ICNGGVEP---------G 467
RGCFEXT    -------DMETELLRF-------RGTPIHFP--GGRP-T--SADAKCWFAQGK-ICQGGIDPALAMMVCF-A 475
RGCGEXT    ------KVIRP---WR-------DDLNASGP--HGSH-P--EYKPDCGFHED--LCRTKP-PTGAGMT---A 477
SPGCEXT    G--TFNRENGQWGFEEALDEDANVLKPV-WH--NRDEPP--LDMPPCGFHGELCT----NWALYLGAS---- 514
ruler      .......610.......620.......630.......640.......650.......660.......670..
```

Fig. 8. Sequence alignment of extracellular domains of transmembrane guanylyl cyclases. Sequences encoding the extracellular domain of known transmembrane guanylyl cyclases or sequences identified through database searches that are predicted to encode transmembrane guanylyl cyclases were aligned. Note the conservation of cysteines (colored red)

Function of Specific Cyclases. Specific inhibition of enzymes/receptors has represented one of the most powerful means by which to define the function of a protein. However, in the case of the cyclases, highly specific drug inhibitors have not been reported (various compounds such as ODQ, methylene blue, and HS-142 inhibit multiple forms of the cyclases). One method to inhibit a specific cyclase signaling pathway is disruption of the corresponding gene. Many caveats to the interpretation of function based on gene disruption exist, but in combination with other studies the physiological role of some of the cyclases is more clearly understood. The genes that have been disrupted so far are those for the membrane forms of cyclase (GC-A, GC-C and GC-E). GC-A appears to play an important role in blood pressure regulation since disruption of its gene has recently been shown to result in a salt-resistant form of hypertension (Lopez et al. 1995, Oliver et al, 1998). Additionally, such mice develop a massive cardiac hypertrophy (Oliver et al. 1997, Wedel and Garbers 1998) which does not appear to be accounted for by pressure alone. Responsiveness to ANP but not CNP is lost in the GC-A null mice showing that it, in fact, is the receptor for ANP (Lopez et al, 1997, Kishimoto et al, 1997). Mice lacking the GC-C receptor are resistant to E. coli heat-stable enterotoxins, but the physiological role of GC-C remains obscure (Mann et al. 1997 Schulz et al. 1997), since the null mice display no obvious phenotype different from wild-type mice, except for the resistance to STa. Disruption of the GC-E gene results in a specific loss of cones in null mice (unpublished results), but the rod photoreceptors remain morphologically normal. Recently, mutations in GC-E have been associated with two human genetic visual diseases, Leber's congenital amaurosis and cone-rod dystrophy (Perrault et al. 1996, Kelsell et al. 1998, Perrault et al. 1998). Although the reason for the specificity of the gene defect on cone survival is not yet known, GC-E has been reported to be expressed in both cones and rods, whereas another retinal guanylyl cyclase (GC-F) has been found only in the rod photoreceptors to date (Dizhoor et al. 1994, Liu et al. 1994, Cooper et al. 1996, Yang and Garbers 1997). Thus, the GC-E gene disruption may eliminate a source of cyclic GMP in the cones, leading to their ultimate demise. The generation of mice lacking the other guanylyl cyclase genes will certainly contribute toward our understanding of the complex physiological functions of these molecules.

Conclusions. Although enormous progress in understanding guanylyl cyclase structure and regulation has been recently made, many questions remain. There are now numerous guanylyl cyclase sequences, and signature sequences for discrete domains within these molecules can be identified. The majority of the guanylyl cyclases identified, however are orphan receptors. A primary goal, therefore, is the identification of ligands for the numerous orphan receptor guanylyl cyclases. Identification of these molecules may provide insight into systems such as vision and olfaction among others. In addition, it will be of interest to identify guanylyl cyclase regulatory proteins. These molecules may provide insight into guanylyl cyclase regulation and provide targets for other signaling pathways to modulate guanylyl cyclase activity. Finally, the information gained from structural studies of adenylyl cyclase has shed new light on the guanylyl cyclase catalytic domain, and raised the possibility of a previously unidentified regulatory pocket within the catalytic domain. Understanding the role of this potential regulatory region may provide new insight into not only guanylyl cyclase regulation, but numerous physiological processes.

References

Aparicio JG, Applebury ML (1996) The photoreceptor guanylate cyclase is an autophosphorylating protein kinase. J Biol Chem *271*:27083–27089

Baude EJ, Arora VK, Yu S, Garbers DL, Wedel BJ (1997) The cloning of a Caenorhabditis elegans guanylyl cyclase and the construction of a ligand-sensitive mammalian/nematode chimeric receptor. J Biol Chem *272*:16035–16039

Bentley JK, Tubb DJ, Garbers DL (1986) Receptor-mediated activation of spermatozoan guanylate cyclase. J Biol Chem *261*:14859–14862

Chang C, Kohse KP, Chang B, Hirata M, Jiang B, Douglas JE, Murad F (1990) Characterization of ATP-stimulated guanylate cyclase activation in rat lung membranes. Biochim Biophys Acta *1052*:159–165

Chinkers M, Singh S, Garbers DL (1991) Adenine nucleotides are required for activation of rat atrial natriuretic peptide receptor/guanylyl cyclase expressed in a baculovirus system. J Biol Chem *266*:4088–4093

Chinkers M, Garbers DL (1989) The protein kinase domain of the ANP receptor is required for signaling. Science *245*:1392–1394

Chinkers M, Wilson EM (1992) Ligand-independent oligomerization of natriuretic peptide receptors:Identification of heteromeric receptors and a dominant negative mutant. J Biol Chem *267*:18589–18597

Cooper N, Liu L, Yoshida A, Pozdnyakov N, Margulis A, Sitaramayya A (1996) The bovine rod outer segment guanylate cyclase, ROS-GC1, is present in both outer and synaptic layers of the retina. J Mol Neurosci *6*:211–222

Crane JK, Shanks KL (1996) Phosphorylation and activation of the intestinal guanylyl cyclase receptor for Escherichia coli heat-stable toxin by protein kinase C. Mol Cell Biochem *165*:111–120

Dessauer CW, Gilman AG (1997) The catalytic mechanism of mammalian adenylyl cyclase. Equilibrium binding and kinetic analysis of P-site inhibition. J Biol Chem *272*:27787–27795

Dizhoor AM, Lowe DG, Olshevskaya EV, Laura RP, Hurley JB (1994) The human photoreceptor membrane guanylyl cyclase, RetGC, is present in outer segments and is regulated by calcium and a soluble activator. Neuron *12*:1345–1352

Drewett JG, Garbers DL (1994) The family of guanylyl cyclase receptors:Their ligands and functions. Endocrine Reviews *15*:135–162

Fenrick R, McNicoll N, De Lean A (1996) Glycosylation is critical for natriuretic peptide receptor-B function. Mol Cell Biochem *165*:103–109

Foerster J, Harteneck C, Malkewitz J, Schultz G, Koesling D (1996) A functional heme-binding site of soluble guanylyl cyclase requires intact N-termini of alpha 1 and beta 1 subunits. Eur J Biochem *240*:380–386

Foster DC, Garbers DL (1998) Dual role for adenine nucleotides in the regulation of the atrial natriuretic peptide receptor, guanylyl cyclase-A. J Biol Chem *273*:16311–16318

Friebe A, Schultz G, Koesling D (1996) Sensitizing soluble guanylyl cyclase to become a highly CO-sensitive enzyme. EMBO J *15*:6863–6868

Friebe A, Koesling D (1998) Mechanism of YC-1-induced activation of soluble guanylyl cyclase. Mol Pharmacol 53:123–127

Garbers DL (1979) Purification of soluble guanylate cyclase from rat lung. J Biol Chem *254*:240–243

Garbers DL, Koesling D, Schultz G (1994) Guanylyl cyclase receptors. Mol Biol Cell *5*:1–5

Garbers DL, Lowe DG (1994) Guanylyl cyclase receptors. J Biol Chem *269*:30741–30744

Garthwaite J, Southam E, Boulton CL, Nielsen EB, Schmidt K, Mayer B (1995) Potent and selective inhibition of nitric oxide-sensitive guanylyl cyclase by 1H-[1,2,4]oxadiazolo[4,3-a]quinoxalin-1-one. Mol Pharmacol *48*:184–188

Gerzer R, Böhme E, Hofmann F, Schultz G (1981) Soluble guanylate cyclase purified from bovine lung contains heme and copper. FEBS Lett *132*:71–74

Goraczniak RM, Duda T, Sharma RK (1992) A structural motif that defines the ATP-regulatory module of guanylate cyclase in atrial natriuretic factor signalling. Biochem J *282*:533–537

Gow AJ, Stamler JS (1998) Reactions between nitric oxide and haemoglobin under physiological conditions. Nature *391*:169–173

Harteneck C, Koesling D, Soling A, Schultz G, Bohme E (1990) Expression of soluble guanylate cyclase:catalytic activity requires two enzyme subunits. FEBS Lett *272*:221–223

Hobbs AJ (1997) Soluble guanylate cyclase:the forgotten sibling. Trends Pharmacol Sci *18*:484–491

Humbert P, Niroomand F, Fischer G, Mayer B, Koesling D, Hinsch K, Gausepohl H, Frank R, Schultz G, Böhme E (1990) Purification of soluble guanylyl cyclase from bovine lung by a new immunoaffinity chromatographic method. Eur J Biochem *190*:273–278

Ignarro LJ, Wood KS, Wolin MS (1982) Activation of purified soluble guanylate cyclase by protoporphyrin IX. Proc Natl Acad Sci USA *79*:2870–2873

Ignarro LJ, Wood KS, Wolin MS (1984) Regulation of purified soluble guanylate cyclase by porphyrins and metalloporphyrins:a unifying concept. Adv Cyclic Nucleo Res *17*:267–274

Itakura M, Iwashina M, Mizuno T, Ito T, Hagiwara H, Hirose S (1994) Mutational analysis of disulfide bridges in the type C atrial natriuretic peptide receptor. J Biol Chem *269*:8314–8318

Kamisaki Y, Saheki S, Nakane M, Palmieri JA, Kuno T, Chang BY, Waldman SA, Murad F (1986) Soluble guanylate cyclase from rat lung exists as a heterodimer. J Biol Chem *261*:7236–7241

Kelsell RE, Gregory-Evans K, Payne AM, Perrault I., Kaplan J., Yang RB, Garbers DL, Bird AC, Moore AT, Hunt DM (1998) Mutations in the retinal guanylate cyclase (RETGC-1) gene in dominant cone-rod dystrophy. Hum Mol Genet *7*:1179–1184

Kharitonov VG, Sharma VS, Pilz RB, Magde D, Koesling D (1995) Basis of guanylate cyclase activation by carbon monoxide. Proc Natl Acad Sci USA *92*:2568–2571

Kishimoto I, Dubois SK, Garbers DL (1996) The heart communicates with the kidney exclusively through the guanylyl cyclase-A receptor:acute handling of sodium and water in response to volume expansion. Proc Natl Acad Sci USA *93*:6215–6219

Koller KJ, Lipari MT, Goeddel DV (1993) Proper glycosylation and phosphorylation of the type A natriuretic peptide receptor are required for hormone-stimulated guanylyl cyclase activity. J Biol Chem *268*:5997–6003

Kurose H, Inagami T, Ui M (1987) Participation of adenosine 5'-triphosphate in the activation of membrane-bound guanylate cyclase by the atrial natriuretic factor. FEBS Letts *219*:375–379

Landschulz WH, Johnson PF, McKnight SL (1988) The leucine zipper:a hypothetical structure common to a new class of DNA binding proteins. Science *240*:1759–1764

Laura RP, Dizhoor AM, Hurley JB (1996) The membrane guanylyl cyclase, retinal guanylyl cyclase-1, is activated through its intracellular domain. J Biol Chem *271*:11646–11651

Liu X, Seno K, Nishizawa Y, Hayashi F, Yamazaki A, Matsumoto H, Wakabayashi T, Usukura J (1994) Ultrastructural localization of retinal guanylate cyclase in human and monkey retinas. Exp Eye Res *59*:761–768

Liu Y, Ruoho AE, Rao VD, Hurley JH (1997) Catalytic mechanism of the adenylyl and guanylyl cyclases:modeling and mutational analysis. Proc Natl Acad Sci USA *94*:13414–13419

Lopez MJ, Wong SK, Kishimoto I, Dubois S, Mach V, Friesen J, Garbers DL, Beuve A (1995) Salt-resistant hypertension in mice lacking the guanylyl cyclase-A receptor for atrial natriuretic peptide. Nature *378*:65–68

Lopez MJ, Garbers DL, Kuhn M (1997) The guanylyl cyclase-deficient mouse defines differential pathways of natriuretic peptide signaling. J Biol Chem *272*:23064–23068

Lowe DG (1992) Human natriuretic peptide receptor-A guanylyl cyclase is self-associated prior to hormone binding. Biochemistry *31*:10421–10425

Mann EA, Jump ML, Wu J, Yee E, Giannella RA (1997) Mice lacking the guanylyl cyclase C receptor are resistant to STa-induced intestinal secretion. Biochem Biophys Res Commun *239*:463–466

Nakane M, Arai K, Saheki S, Kuno T, Buechler W, Murad F (1990) Molecular cloning and expression of cDNAs coding for soluble guanylate cyclase from rat lung. J Biol Chem *265*:16841–16845

Oliver PM, Fox JE, Kim R, Rockman HA, Kim HS, Reddick RL, Pandey KN, Milgram SL, Smithies O, Maeda N (1997) Hypertension, cardiac hypertrophy, and sudden death in mice lacking natriuretic peptide receptor A. Proc Natl Acad Sci USA *94*:14730–14735

Oliver PM, John SW, Purdy KE, Kim R, Maeda N, Goy MF, Smithies O (1998) Natriuretic peptide receptor 1 expression influences blood pressures of mice in a dose-dependent manner. Proc Natl Acad Sci USA *95*:2547–2551

Perrault I, Rozet JM, Calvas P, Gerber S, Camuzat A, Dollfus H, Chatelin S, Souied E, Ghazi I, Leowski C, Bonnemaison M, Le Paslier D, Frezal J, Dufier JL, Pittler S,

Munnich A, Kaplan J (1996) Retinal-specific guanylate cyclase gene mutations in Leber's congenital amaurosis. Nature Genet *14*:461–464

Perrault I, Rozet JM, Gerber S, Kelsell RE, Souied E, Cabot A, Hunt DM, Munnich A, Kaplan J (1998) A retGC-1 mutation in autosomal dominant cone-rod dystrophy. Am J Hum Genet *63*:651–654

Potter LR (1998) Phosphorylation-dependent regulation of the guanylyl cyclase-linked natriuretic peptide receptor B:dephosphorylation is a mechanism of desensitization. Biochemistry *37*:2422–2429

Potter LR, Garbers DL (1992) Dephosphorylation of the guanylyl cyclase-A receptor causes desensitization. J Biol Chem *267*:14531–14534

Potter LR, Hunter T (1998) Phosphorylation of the kinase homology domain is essential for activation of the A-type natriuretic peptide receptor. Mol Cell Biol *18*:2164–2172

Schrammel A, Behrends S, Schmidt K, Koesling D, Mayer B (1996) Characterization of 1H-[1,2,4]oxadiazolo[4,3-a]quinoxalin-1-one as a heme-site inhibitor of nitric oxide-sensitive guanylyl cyclase. Mol Pharmacol *50*:1–5

Schulz S, Lopez MJ, Kuhn M, Garbers DL (1997) Disruption of the guanylyl cyclase-C gene leads to a paradoxical phenotype of viable but heat-stable enterotoxin-resistant mice. J Clin Invest *100*:1590–1595

Schulz S, Wedel BJ, Matthews A, Garbers DL (1998) The cloning and expression of a new guanylyl cyclase orphan receptor. J Biol Chem *273*:1032–1037

Snyder SH (1992) Nitric oxide:First in a new class of neurotransmitters? Science *257*:494–496

Stone JR, Marletta MA (1995a) Heme stoichiometry of heterodimeric soluble guanylate cyclase. Biochemistry *34*:14668–14674

Stone JR, Marletta MA (1995b) The ferrous heme of soluble guanylate cyclase:formation of hexacoordinate complexes with carbon monoxide and nitrosomethane. Biochemistry *34*:16397–16403

Stone JR, Marletta MA (1996) Spectral and kinetic studies on the activation of soluble guanylate cyclase by nitric oxide. Biochemistry *35*:1093–1099

Stone JR, Marletta MA (1998) Synergistic activation of soluble guanylate cyclase by YC-1 and carbon monoxide:implications for the role of cleavage of the iron-histidine bond during activation by nitric oxide. Chem Biol *5*:255–261

Stults JT, O'Connell KL, Garcia C, Wong S, Engel AM, Garbers DL, Lowe DG (1994) The disulfide linkages and glycosylation sites of the human natriuretic peptide receptor-C homodimer. Biochemistry *33*:11372–11381

Sunahara RK, Beuve A, Tesmer JJ, Sprang SR, Garbers DL, Gilman AG (1998) Exchange of substrate and inhibitor specificities between adenylyl and guanylyl cyclases. J Biol Chem *273*:16332–16338

Tang WJ, Gilman AG (1995) Construction of a soluble adenylyl cyclase activated by Gs alpha and forskolin. Science *268*:1769–1772

Taylor SS, Knighton DR, Zheng J, Ten Eyck LF, Sowadski JM (1992) Structural framework for the protein kinase family. Ann Rev Cell Biol *8*:429–462

Tesmer JJ, Sunahara RK, Gilman AG, Sprang SR (1997) Crystal structure of the catalytic domains of adenylyl cyclase in a complex with Gsalpha.GTPgammaS. Science *278*:1907–1916

Thompson DK, Garbers DL (1995) Dominant negative mutations of the guanylyl cyclase-A receptor. Extracellular domain deletion and catalytic domain point mutations. J Biol Chem *270*:425–430

Thorpe DS, Morkin E (1990) The carboxyl region contains the catalytic domain of the membrane form of guanylate cyclase. J Biol Chem *265*:14717–14720
Traylor TG, Sharma VS (1992) Why NO?. Biochemistry *31*:2847–2849
Tucker CL, Hurley JH, Miller TR, Hurley JB (1998) Two amino acid substitutions convert a guanylyl cyclase, RetGC-1, into an adenylyl cyclase. Proc Natl Acad Sci USA *95*:5993–5997
Vaandrager AB, van der Wiel E, de Jonge HR (1993) Heat-stable enterotoxin activation of immunopurified guanylyl cyclase C. Modulation by adenine nucleotides. J Biol Chem *268*:19598–19603
Verma A, Hirsch DJ, Glatt CE, Ronnett GV, Snyder SH (1993) Carbon monoxide:a putative neural messenger. Science *259*:381–384
Wada A, Hasegawa M, Matsumoto K, Niidome T, Kawano Y, Hidaka Y, Padilla PI, Kurazono H, Shimonishi Y, Hirayama T (1996) The significance of Ser1029 of the heat-stable enterotoxin receptor (STaR):relation of STa-mediated guanylyl cyclase activation and signaling by phorbol myristate acetate. FEBS Lett *384*:75–77
Wedel BJ, Garbers DL (1998) Guanylyl cyclases:approaching year thirty. Trends Endocinol Metab *9*:213–219
Wedel B, Humbert P, Harteneck C, Foerster J, Malkewitz J, Bohme E, Schultz G, Koesling D (1994) Mutation of His-105 in the beta 1 subunit yields a nitric oxide-insensitive form of soluble guanylyl cyclase. Proc Natl Acad Sci USA *91*:2592–2596
Wedel B, Harteneck C, Foerster J, Friebe A, Schultz G, Koesling D (1995) Functional domains of soluble guanylyl cyclase. J Biol Chem *270*:24871–24875
Weikel CS, Spann CL, Chambers CP, Crane JK, Linden J, Hewlett EL (1990) Phorbol esters enhance the cyclic GMP response of T84 cells to the heat-stable enterotoxin of *Escherichia coli* (STa). Infection and Immunity *58*:1402–1407
Yang R, Foster DC, Garbers DL, Fülle H (1995) Two membrane forms of guanylyl cyclase found in the eye. Proc Natl Acad Sci USA *92*:602–606
Yang RB, Garbers DL (1997) Two eye guanylyl cyclases are expressed in the same photoreceptor cells and form homomers in preference to heteromers. J Biol Chem *272*:13738–13742
Zhang G, Liu Y, Ruoho AE, Hurley JH (1997) Structure of the adenylyl cyclase catalytic core. Nature *386*:247–253
Zhao Y, Marletta MA (1997) Localization of the heme binding region in soluble guanylate cyclase. Biochemistry *36*:15959–15964
Zhao Y, Schelvis JPM, Babcock GT, Marletta MA (1998) Identification of histidine 105 in the β1 subunit of soluble guanylate cyclase as the heme proximal ligand. Biochemistry *37*:4502–4509

Appendix

A2, A3, A4, A8:rat AC; A5, A6:Canis familiaris AC; ac7:bovine AC; AC9:Mus musculus AC; HSGCSA-B:human sGC; RGA1, RGSA2, RGB1, RGB2:rat sGC; RGCA-G:rat mGC; BSB1:bovine sGC; PIGSTA:pig GC-C; FGCC:Xenopus laevis GC-C; DMU117, DMU123:Drosophila melanogaster sGC; DRG0-1:Drosophila melanogaster mGC; drorac:Drosophila melanogaster AC; GCX1, GCY1-33, M04G12, B0024, T04D3, F52E1, ZK896, ZC239, ZC412, F21H7, C30G4:Caenorhabditis elegans GC; cef17:Caenorhabditis elegans AC;

OLG3-5, OLGA-B:Oryzias latipes GC; SPGC:Strogylocentrotus purpuratus GC; EGC-B:Anguilla japonica GC-B; CION:Ciona intestinalis GC; AMU18, AMU20:anopheles sGC; CFYX83, CFY484:Canis familiaris GC.

Soluble Guanylyl Cyclase: Structure and Regulation

D. Koesling and A. Friebe

Institut für Pharmakologie, Freie Universität Berlin, Thielallee 69–73, D-14195 Berlin, Germany

Contents

1 Introduction

Soon after the discovery of cAMP, cGMP was detected in urine (Ashman et al. 1963), and, later on, was shown to be formed in practically all cells and tissues (Hardman and Sutherland, 1969; Schultz et al. 1969; White and Auerbach, 1969; Goldberg et al. 1973). Although intracellular concentrations are lower than those of cAMP, cGMP is an important signaling molecule involved in synaptic transmission (O'Dell et al. 1991; Zhuo and Hawkins, 1995), smooth muscle relaxation (Lincoln, 1989), inhibition of platelet aggregation and retinal signal transduction (Garthwaite et al. 1988; Walter, 1989; Schmidt and Walter, 1994; Moncada and Higgs, 1995). The effects of cGMP are mediated by cGMP-dependent protein kinases, cGMP-gated ion channels and cGMP-regulated phosphodiesterases (for review see other chapters in this issue). In contrast to the cAMP-forming adenylyl cyclases which are restricted to the plasma membrane in eukaryotic cells with the exception of the cytosolic adenylyl cyclase in sperm (Braun and Dods, 1975), the cGMP-synthesizing guanylyl cyclases exist as soluble and membrane-bound forms (Fig. 1). Although in the early days of guanylyl cyclase research considerable effort was made regarding the characterization and isolation of the two enzyme forms, major progress was not achieved until the identification of the activators of the guanylyl cyclases, i.e. nitric oxide for the soluble enzyme (Palmer et al. 1987) and natriuretic peptides for the membrane forms (Chinkers et al. 1989). Whereas natriuretic peptides represent a group of peptide hormones which stimulate several isoforms of membrane-bound guanylyl cyclases, nitric oxide is the only physiologically occurring activator of the two isoforms of the soluble enzyme known to date. The membrane-bound guanylyl cyclases and their respective activators are discussed in more detail in another chapter of this issue.

The research about biological effects of nitric oxide and about the structure and regulation of the nitric oxide-forming enzymes, i.e. the NO synthases, has expanded into an own field of investigation. Most of the functions of NO as a signaling molecule are mediated by the stimulation of soluble guanylyl cyclase (sGC). This chapter will give an overview on sGC emphasizing recent findings about new modulators of the enzyme.

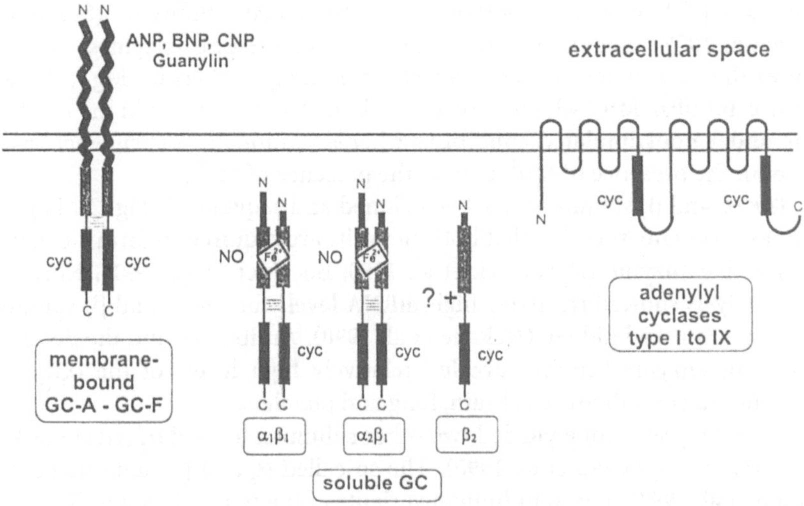

Fig. 1. Schematic representation of the overall structures of guanylyl cyclases and adenylyl cyclases. Shown are the different heterodimers of soluble guanylyl cyclase ($\alpha_1\beta_1$, $\alpha_2\beta_1$), a representative membrane-bound guanylyl cyclase with different ligands at the respective receptor domain and a representative membrane-bound adenylyl cyclase. The dimerising partner of the β_2 subunit of guanylyl cyclase has not been identified yet. Boxes labeled with cyc represent the cyclase catalytic domains homologous in the guanylyl and adenylyl cyclases. ANP, natriuretic peptide type A; BNP, natriuretic peptide type B; CNP, natriuretic peptide type C; cyc, putative catalytic domain of guanylyl cyclases and adenylyl cyclases; GC, guanylyl cyclase; GC-A - GC-F, membrane-bound guanylyl cyclase type A to type F

2 Enzyme Structure

2.1 Isoforms and Tissue Distribution

sGC has been successfully purified by several groups, most of them using lung tissue as starting material. All isolation procedures yielded an enzyme consisting of two subunits termed α_1 and β_1 (see Fig. 1). Whereas the β_1 subunit of the enzyme purified by several groups always revealed a molecular mass of 70 kDa, the molecular mass reported for the α_1 subunit differed between 73 to 80 kDa (Gerzer et al. 1981; Humbert et al. 1990; Stone and Marletta, 1994; Burstyn et al. 1995; Tomita et al. 1997). The purified enzyme was stimulated up to 400-fold by NO and was shown to contain a prosthetic heme group (Gerzer et al. 1981). Specific activities of 25 to 40 μmole cGMP x

$min^{-1} \times mg^{-1}$ have been reported under stimulated conditions (Stone and Marletta, 1995). Similar to other nucleotide-converting enzymes, sGC requires divalent metal ions as a cofactor for catalysis. Besides Mg^{2+}, the enzyme can utilize Mn^{2+} which leads to an about 4-fold increase in the catalytic rate under nonstimulated conditions whereas stimulated enzyme activity is lower in the presence of Mn^{2+} than in the presence of Mg^{2+}.

The α_1 and β_1 subunits have been cloned and sequenced (Fig. 2). Expression experiments revealed that both subunits are required to form a catalytically active enzyme (Harteneck et al. 1990; Buechler et al. 1991). Northern blot analysis showed relatively high mRNA levels for the α_1 and β_1 subunits in lung, brain and kidney (Nakane et al. 1990). Antibodies and the determination of enzyme activity revealed relatively high levels of the $\alpha_1\beta_1$ heterodimer in smooth muscle, brain, lung and platelets.

Homology screening yielded two other subunits, termed α_2 (Harteneck et al. 1991) and β_2 (Yuen et al. 1990). The so-called α_3 and β_3 subunits of sGC (Giuili et al. 1992) represent human variants of the α_1 and β_1 subunits rather than different isoforms, and changes in the reading frame probably account

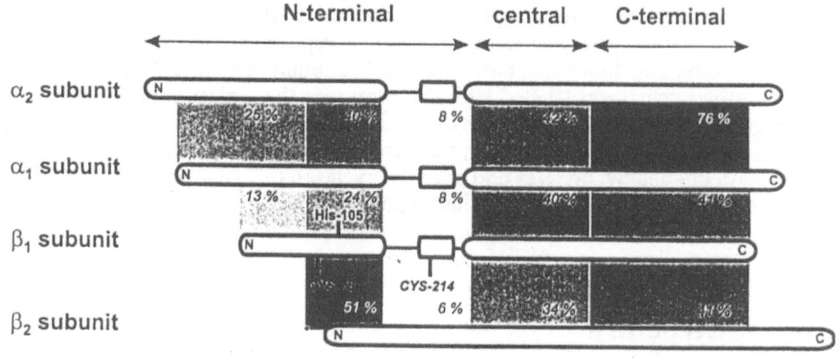

Fig. 2. Comparison of the subunits of sGC. The polypeptide chains are shown as white bars, gaps are indicated by a black line. The amount of identical amino acids shared between the respective subunits in a certain region is symbolized by the shaded box between the subunits and also given in percent. The C-termini of the subunits of sGC are unrelated. The position of the His-105 as the heme-coordinating residue of the β_1 subunit is given; in addition, the positions of the cysteines (Cys-78 and Cys-214 of the β_1 subunit) whose mutation resulted in a loss of the prosthetic heme group, and of their counterparts in the α_1 subunit (Cys-145 and Cys-284) whose mutation did not affect heme-binding are given. The regions chosen for comparison are: amino acids 1-161 of the α_2, 1-126 of the α_1, and 1-56 of the β_1 subunits; 162-268 of the α_2, 127-234 of the α_1, 57-164 of the α_2, and 1-101 of the β_2 subunits; 269-321 of the α_2, 235-282 of the α_1, 165-212 of the β_1, and 102-208 of the β_2 subunits; 322-483 of the α_2, 283-441 of the α_1, 213-383 of the β_1, and 205-368 of the β_2 subunits; 484-698 of the α_2, 442-660 of the α_1, 384-605 of the β_1, and 369-581 of the β_2 subunits

for the differences in amino acid sequence. In most laboratories, the β_2 subunit has not formed a cGMP-synthesizing enzyme with any of the known subunits in expression experiments. Yet, there is a recent report on the formation of an active $\alpha_1\beta_2$ heterodimer (Gupta et al. 1997). In contrast to β_2, the α_2 subunit is able to substitute for the α_1 subunit as shown by the formation of a catalytically active, NO-sensitive heterodimer upon coexpression with the β_1 subunit. A catalytically inactive splice-variant of the α_2 subunit has been identified (Behrends et al. 1995). In expression systems, this subunit was still able to form heterodimers with the β_1 subunit and was able to inhibit the formation of the catalytically active $\alpha_2\beta_1$ heterodimer, thereby acting as a dominant negative subunit. PCR analysis revealed an almost ubiquitous distribution of α_1, α_2 and β_1 subunits and showed mRNA for the splice-variant in some but not all of the α_2 containing tissues (Behrends et al. 1995).

Recently, the natural occurrence of the α_2 subunit has been demonstrated in human placenta with the help of precipitating antibodies. As the β_1 subunit coprecipitated with the α_2 subunit, the β_1 subunit appears to be the corresponding dimerization partner of the α_2 subunit under physiological conditions (Russwurm et al. 1998). Analysis of the purified $\alpha_2\beta_1$ heterodimer did not reveal any differences to the $\alpha_1\beta_1$ heterodimer with respect to the heme content, the sensitivity towards NO, the kinetic properties and the responsiveness towards modulators. The lack of differences between the isoforms is surprising taking into consideration that the N-terminal thirds of the α subunits share only 27% identical amino acids. Right now, the physiological relevance of the two so far indistinguishable isoforms is unclear. Expression under the transcriptional control of cell type-specific or developmental promoters may be a plausible explanation for the existence of two functionally similar isoforms.

sGC subunits have also been found in animals other than mammals. In Drosophila melanogaster, analogs to the α_1 and β_1 subunits of sGC have been identified (Yoshikawa et al. 1993; Shah and Hyde, 1995). Expression of both subunits yielded an NO-sensitive guanylyl cyclase. In the medaka fish, Oryzias latipes, subunits corresponding to mammalian α_1 and β_1 were detected (Mikami et al. 1998). In Caenorhabditis elegans there are at least four sequences exhibiting homology to the subunits of sGC, but an unequivocal classification into α or β subunits is not obvious (Yu et al. 1997).

2.2 Primary Structure and Homology Among the Subunits of sGC

Four subunits of sGC (α_1, α_2, β_1, β_2) are known to date (Fig. 2). The cDNAs of the α_1 and β_1 subunits were identified with the help of peptide sequences derived from the enzyme purified from rat and bovine lung (for review see Koesling et al. 1991). The cDNAs of the α_2 (Harteneck et al. 1991) and β_2 (Yuen et al. 1990) subunits were identified by homology screening. A comparison of the primary structure of the subunits of sGC reveals that they can be divided in three parts: a C-terminal cyclase catalytic domain, a central part and an N-terminal region (see Fig. 2). The highest degree of homology is found in the C-terminal catalytic domains of sGC which are also similar to the respective regions in the membrane-bound guanylyl cyclases and in the adenylyl cyclases (see also under 'Catalytic domain'). The C-terminus of the β_2 subunit extends 86 amino acids beyond the C-terminus of the β_1 subunit and contains an isoprenylation consensus sequence (-CVVL); therefore, attachment of the β_2 subunit to the membrane via an isoprenoid group appears possible.

The catalytic domains are preceded by a region with considerable homology among the subunits. In analogy to the membrane-bound enzymes (Wilson and Chinkers, 1995), these regions are probably involved in the dimerization of the subunits.

The N-terminal regions of the subunits are relatively diverse. For example, only 27% identical amino acids are shared by the otherwise highly conserved α_1 and α_2 subunits. The differences in primary structure of both α subunits are contrasted by their functional similarity (see above under 'Isoforms') indicating that most of the regulatory features of sGC are determined by the β_1 subunit. Within the N-terminal regions, there is a stretch of about 100 amino acids which shows a higher degree of conservation among the α subunits or the β subunits than between α and β. Conceivably, these regions define the properties of an α or a β subunit. It should be noted that the histidine 105 of the β_1 subunit (see below) which was identified as a candidate for the proximal ligand of the heme group is located within this region.

2.3 Catalytic Domain

All subunits of sGC, the membrane-bound GCs and the adenylyl cyclases contain a conserved C-terminal domain of about 250 amino acids. Adenylyl cyclases, of which 9 different isoforms have been identified to date, are integral membrane proteins with two hydrophobic domains, each hypothesized to contain six transmembrane helices, and two 40 kDa cytosolic domains termed C1 and C2 (Sunahara et al. 1996). The cytosolic domains are homologous to each other, conserved among the adenylyl cyclases and homologous to the catalytic domains of membrane-bound GCs and the subunits of sGC. Like sGC, adenylyl cyclases contain two different, although conserved catalytic domains. Yet, in the adenylyl cyclases, both domains are located on one molecule whereas in sGC each of the two subunits contributes one catalytic domain. Comparison of the primary structure of the catalytic domains of adenylyl cyclases and the subunits of sGC suggests that the catalytic domains of the α and β subunits correspond to the C1 and C2 domains of the adenylyl cyclases, respectively. The structural similarity is underlined by the finding that, at least under stimulated conditions, sGC is able to use ATP as a substrate and to catalyze its conversion into cAMP (Mittal and Murad, 1977).

Evidence that these domains indeed represent the catalytic domains was brought by deletion mutants of all three cyclic nucleotide-forming enzymes (Wedel et al. 1995; Thompson and Garbers, 1995; Tang et al. 1995). All deletion mutants which consisted mainly of the conserved C-terminal regions were still catalytically active. Apparently, the common cyclase catalytic domain has been conserved during evolution and has been linked to different regulatory modules such as the heme-binding domain in the sGCs, a receptor domain in the membrane-bound GCs, and G-protein-interacting domains in the adenylyl cyclases.

It has to be pointed out that dimer formation is a prerequisite for catalytic activity for all cyclases. Coexpression of an α and β subunit or of the truncated α_1 and β_1 subunits was required to yield a catalytically active sGC (Wedel et al. 1995). The membrane-bound guanylyl cyclases also consist of at least two subunits, and the truncated, still catalytically active membrane-bound guanylyl cyclase has a homodimeric structure (Thompson and Garbers, 1995). Analogous results exist for the catalytic domains of the adenylyl cyclases where coexpression of the C1 and C2 domains yields an enzymatically active enzyme (Tang et al. 1995; Dessauer and Gilman, 1996). Although the C2 domain is able to form homodimers, its cAMP-forming activity is negligible.

Recent analysis of the crystal structure of the catalytic domains of adenylyl cyclase showed that the binding site for ATP is formed by both the C1 and C2 domains and suggested a few amino acid residues to be responsible for nucleotide interaction (Tesmer et al. 1997). Indeed, the exchange of the corresponding residues of sGC to their equivalents in the adenlyly cyclases resulted in an sGC forming cAMP in an NO-sensitive manner (Sunahara et al. 1998). These data underline the structural similarity among the adenylyl and soluble guanylyl cyclases.

2.4
The Regulatory Heme-Binding Domain

sGC has been purified as heme-deficient and as heme-containing enzymes (Gerzer et al. 1981; Ohlstein et al. 1982) The presence of a prosthetic heme group is a prerequisite for the stimulatory effect of NO (Craven and DeRubertis, 1978; Ignarro et al. 1982). Removal of the heme group abolishes NO-induced activation, and reconstitution of sGC with the heme group restores NO sensitivity (Ignarro et al. 1986; Foerster et al. 1996). The heme content of the enzyme has been a matter of debate. Whereas most groups found up to 1 mole of heme per mole of heterodimer (Gerzer et al. 1981; Humbert et al. 1990; Tomita et al. 1997), the group of Marletta reported a heme content of 1.5 mole of heme per mole of heterodimer (Stone and Marletta, 1995). However, recent results obtained in that group are consistent with a heme to heterodimer ratio of 1 : 1 (Stone and Marletta, 1998).

Truncated enzymes, in which the poorly conserved N-terminal 134 and 64 amino acids of the α_1 and β_1 subunits had been deleted, showed impaired or improper heme-binding (Foerster et al. 1996). The truncation of the α subunit resulted in an enzyme with a greatly reduced heme-binding capacity. This truncated mutant still exhibited some sensitivity to NO indicating proper coordination of the remaining heme. In contrast, deletion of the 64 N-terminal amino acids of the β_1 subunit resulted in an enzyme that was still able to bind heme in amounts comparable to the wild type enzyme. However, the absorption maximum for the heme was shifted indicating improper coordination of the heme. In agreement with these findings, this heme-containing mutant was not stimulated by NO. Thus, these results show that the N-termini of the sGC subunits are involved in heme-binding. It appears that the α_1 and β_1 subunits contribute unequally to the heme-binding. This notion is supported by cysteine mutants originally created to study a possible redox regulation of sGC (Friebe et al. 1997). Mutation of two cysteines (Cys-78 and Cys-214) located in the N-terminal, heme-binding region of the β_1 subunit (see Fig. 2) yielded proteins that were insensitive to NO due to loss of enzyme-bound heme. Both mutants could be reconstituted with heme and, as a consequence, regained NO sensitivity which is an indicator for intact intramolecular signaling. Apparently, the mutation of the two cysteines of the β_1 subunit resulted in a reduced affinity of sGC for heme. Mutation of the corresponding cysteines on the α_1 subunit (see Fig. 2) did not alter NO responsiveness, underlining the exclusive role of the β_1 subunit in heme binding. Moreover, in a recent report, the N-terminal half of the β_1 subunit (amino acid 1-385) has been shown to be sufficient to bind heme in a manner similar to the wild type enzyme as shown by spectral analysis (Zhao and Marletta, 1997).

The absorption maximum of the heme group of sGC (431 nm) is indicative of a histidine as the axial ligand. In order to identify the histidine responsible for heme-binding, conserved histidines of both subunits were point-mutated to phenylalanine. Analysis of the respective mutants revealed that the histidine 105 of the β_1 subunit is essential for NO stimulation (Wedel et al. 1995). In contrast to all other histidine mutants, the His-105 mutant was insensitive to NO but exhibited intact basal activity. Since the His-105 is preceded by a leucine in the -4 position, similar to the proximal histidine of all hemo- and myoglobin chains (Fig. 3), the His-105 is a good candidate for the heme-binding residue in sGC. Recently, the identity of the His-105 as the heme-binding residue on the β_1 subunit was confirmed by Zhao et al. (1998).

α hemoglobin chain	L	S	A	L	S	D	L	H	A	man
	L	S	E	L	S	D	L	H	A	cow
	L	S	A	L	S	D	L	H	A	platypus
	L	S	K	L	S	D	L	H	A	chicken
β hemoglobin chain	F	A	T	L	S	E	L	H	C	man
	F	A	A	L	S	E	L	H	C	cow
	F	A	K	L	S	E	L	H	C	platypus
	F	S	Q	L	S	E	L	H	C	chicken
myoglobin	I	K	P	L	A	Q	S	H	A	man
	V	K	H	L	A	E	S	H	A	cow
	L	K	P	L	A	Q	S	H	A	platypus
	L	K	P	L	A	Q	T	H	A	chicken
soluble guanylyl cyclase	K	A	T	L	E	Q	A	H	Q	α_1 subunit
	K	A	T	L	E	R	T	H	Q	α_2 subunit
	L	Q	N	L	D	A	L	H	D	β_1 subunit
	I	E	N	L	D	A	L	H	S	β_2 subunit

Fig. 3. Vicinities of the proximal histidines in different hemo- and myoglobin chains. Shown are the amino acids preceding the proximal histidine of the α and β hemo- and myoglobin chains of the indicated species. Marked are the proximal histidines and the leucines in the –4 position, which are conserved in all vertebrate hemo- and myoglobin types. The last two lines show the amino acids of the β_1 (99-106) and β_2 subunits (36–44) with the putative proximal histidine 105 of the β_1 subunit which is conserved in the β_2 subunit. For further explanations see text

The His-105 mutant did not contain any detectable heme after purification showing that this histidine contributes to heme-binding. In order to find out whether the lack of the heme group is the only reason for the observed NO insensitivity, heme reconstitution experiments were performed. These experiments showed that, despite the amino acid exchange, the His-105 mutant was still able to bind heme (Foerster et al. 1996). Yet, the Soret absorption of the enzyme bound heme was shifted from 431 nm to 400 nm. This shift is compatible with the altered axial coordination of the heme iron expected for a mutant enzyme lacking the axial heme coordinating residue. Despite the presence of the prosthetic heme group, the reconstituted His-105 mutant remained insensitive to NO. The non-responsiveness to NO of the reconstituted mutant suggests that the His-105, besides being involved in heme-binding, is required to transduce the NO-induced, heme-mediated activation within the protein.

3 Regulation

3.1 NO-Dependent Activation

Although in the late seventies NO-containing compounds were shown to be potent activators of sGC (Arnold et al. 1977; Böhme et al. 1978), the physiological significance of NO-induced activation of the enzyme did not become clear until the endothelium-derived relaxing factor (EDRF) was identified as NO (Palmer et al. 1987, Ignarro et al. 1987). EDRF had been shown to be synthesized in endothelial cells in response to vasodilatory agonists such as acetylcholine, histamine and bradykinin, and to cause vasodilation via stimulation of sGC in smooth muscle cells. Soon after the discovery of NO in the vascular system, NO formation was reported to occur throughout the body, and during the last years the research on NO has expanded widely (Moncada and Higgs, 1995). The enzymes responsible for the synthesis of NO were identified and termed NO synthases. Three isoforms of NO synthases are known to date: The inducible isoform plays a role in the nonspecific immune response and produces relatively high NO concentrations that exhibit direct toxic effects. The neuronal and endothelial NO synthases are constitutively expressed and are regulated by the intracellular calcium concentration. These calcium-dependent NO synthases produce comparatively low NO concentrations ranging between 1 and 100 nM. At these low concentrations, NO functions as a signaling molecule, and most of its effects are mediated via the activation of sGC.

Besides being involved in the regulation of smooth muscle tone and inhibition of platelet aggregation, the NO/cGMP signaling cascade has also been shown to participate in neurotransmission in the central and peripheral nervous systems (Garthwaite et al. 1988; O'Dell et al. 1991). Generally, in tissues containing neuronal or endothelial NO synthase, any increase in the intracellular calcium concentration will lead to stimulation of the calcium-sensitive enzymes and eventually to an increased NO production. Due to its high membrane permeability, NO not only stimulates sGC within the NO-producing cell, but also in adjacent cells. A good example to illustrate the NO/cGMP signaling cascade is the endothelium-dependent relaxation mentioned above. Mainly the activation of shear-stress-sensitive channels leads to an increase in the intracellular calcium concentration in the endothelial cell resulting in the activation of the endothelial NO synthase. Following diffusion into the underlying smooth muscle layer, NO stimulates sGC and thereby induces relaxation most likely via stimulation of the cGMP-dependent protein kinase. NO released from the endothelium does not only diffuse into smooth muscle cells but also into the lumen of the blood vessel. There, through the stimulation of sGC in platelets, NO leads to inhibition of platelet aggregation. The antiaggregatory effect of NO is likely to be locally restricted within the lumen of blood vessels as NO is rapidly inactivated by binding to hemoglobin.

3.2
CO-Dependent Activation

CO has been postulated as a physiological activator of sGC despite its rather poor sGC-stimulating properties (see also under 'Mechanism of heme-dependent activation'). There are reports about the role of CO in long-term potentiation (LTP) (Zhou et al. 1993; Stevens and Wang, 1993), olfactory signal transduction (Verma et al. 1993; Leinders-Zufall et al. 1995; Ingi and Ronnett, 1995), vasorelaxation (Utz and Ullrich, 1991; Morita et al. 1995; Zakhary et al. 1996) as well as inhibition of platelet aggregation (Brüne and Ullrich, 1987). CO is enzymatically produced within the body by the heme oxygenases. These enzymes catalyze the degradation of heme to biliverdin and CO (Maines, 1988; Marks et al. 1991) which appears to be a rather costly way for producing a signaling molecule. Besides induction of the CO-producing heme oxygenases, no short term regulation of the enzymes is know so far (Maines, 1997). Thus, although the experiments with the new stimulator of sGC, YC-1, show that, in the presence of YC-1, CO is able to lead to an enormous activation of sGC (Friebe et al. 1996), the debate about the physiological role of CO remains open.

3.3 Mechanism of Heme-Dependent Activation

NO stimulates sGC by binding to the prosthetic heme of the enzyme and leads to an up to 400-fold activation of the purified enzyme (Humbert et al. 1990; Stone and Marletta, 1995). Among the three redox forms of NO (NO^-, $NO^•$, NO^+), only the uncharged NO radical ($NO^•$) was shown to significantly activate the enzyme (Dierks and Burstyn, 1996).

The absorbance spectra of the heme group of sGC exhibits a maximum at 431 nm. This peak indicates a five-coordinated ferrous heme (Fig. 4) with a histidine as the axial ligand at the fifth coordinating position (Stone and Marletta, 1994). Using site-directed mutagenesis of all conserved histidines, His-105 of the β_1 subunit has been identified as candidate for the axial ligand (Wedel et al. 1994, Zhou and Marletta, 1998). Binding of NO to the sixth position of the heme iron leads to the formation of a five-coordinated nitrosyl-heme complex. Hereby, NO pulls the iron out of the plane of the porphyrin ring leading to the breakage of the histidine-to-iron bond. The resulting penta-coordinated nitrosyl-heme exhibits an absorbance maximum at 398 nm (see Fig. 4). The opening of the histidine-to-iron bond is thought to initiate a conformational change resulting in the activation of the enzyme. Yet, the results with the heme-reconstituted His-105 mutant suggest that the opening of the iron bond is not sufficient for activation but that the released His-105 is required to transduce the stimulatory signal.

In accordance with the proposed mechanism of activation, protoporphyrin IX, the iron-free precursor of heme, stimulates sGC independently of NO (Ignarro et al. 1982). As protoporphyrin IX does not contain an iron, its structure resembles that of the NO-heme complex in which the iron is moved out of the plane of the porphyrin ring. In both cases, i.e. in the protoporphyrin IX- and in the NO-stimulated enzyme, the axial histidine is unbound.

For other hemoproteins, the association of NO to a five-coordinate iron is known to be nearly diffusion-controlled whereas the dissociation of NO is generally rather slow. Studies on the dissociation of NO from sGC in the absence of the substrate MgGTP revealed that the enzyme has a comparably high NO dissociation rate. Upon addition of substrate, NO dissociation was increased about 40-fold yielding a half-life of approximately 5 sec at 37 °C for the NO-sGC complex (Kharitonov et al. 1997). This half-life appears fast enough to allow rapid deactivation of the enzyme in biological systems, although it has to be kept in mind that the presence of an NO scavenger is a prerequisite for such fast deactivation.

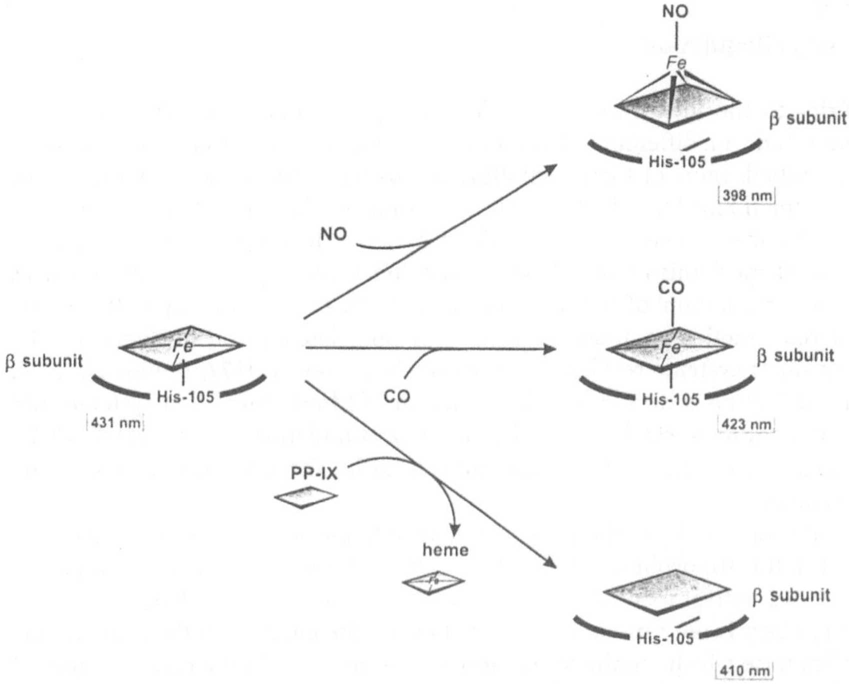

Fig. 4. The heme group of sGC. In the non-activated state, the heme of sGC is five-coordinated with the histidine 105 of the β_1 subunit bound to the fifth coordination position of the heme iron (absorption maximum at 431 nm). After NO binding to the sixth coordination position, the heme is still five-coordinated as the fifth coordination position is now free (absorption maximum at 398 nm). Like NO, CO binds to the sixth coordination position of the heme iron but, in contrast to NO, CO does not lead to a rupture of histidine-to-iron bond; therefore, CO binding results in a six-coordinated heme (absorption maximum at 423 nm). Protoporphyrin IX is able to substitute for the heme group and stimulates the enzyme (absorption maximum at 410 nm). For further explanations see text

Like NO, CO binds to the heme group of sGC with high affinity. In contrast to NO, which activates sGC up to 400-fold, the stimulation achieved by CO only ranges between 4- and 6-fold. Binding of NO induces the breakage of the heme-to-iron bond (see above), whereas CO binding results in a six-coordinated heme. Here, the histidine-to-iron bond remains intact, and the carbonyl-heme complex reveals an absorption maximum at 423 nm (see Fig. 4). Therefore, the low stimulatory property of CO is in accordance with the assumption that the breakage of the histidine-to-iron bond is a prerequisite for sGC activation.

3.5 Redox Regulation

Prior to the identification of NO as the physiological activator of sGC, a multitude of different activators and regulators of the enzyme had been described, most of them exhibiting redox-active properties (for review see Waldman and Murad, 1987). These substances that do not act by releasing NO ('non-NO activators') constitute a rather heterogeneous group; thus, several mechanisms of activation have been postulated and subsumed as 'redòx regulation' of sGC (Böhme et al. 1984). Although many of the effects of redox-active compounds have been postulated to be mediated by the enzyme's cysteine residues (DeRubertis and Craven, 1977; Brandwein et al. 1981; Böhme et al. 1983), point mutation of fifteen conserved cysteines did neither impair catalytic activity nor NO stimulation (Friebe et al. 1997). Thus, the cysteines of sGC are unlikely to be the mediators of this redox regulation.

Purified sGC was shown to be sensitive to the small amounts of NO present in the atmosphere (Friebe et al. 1996a). Studies with one of the reported non-NO activators of sGC, superoxide dismutase, revealed that the stimulatory effect is not caused by a direct effect on the enzyme. Rather, superoxide dismutase which eliminates O_2^- leads to an increase in the concentration of dissolved atmospheric NO in a solution by preventing the inactivation of NO by O_2^- to peroxynitrite (Friebe and Koesling 1998). Other non-NO activators of sGC such as the radical scavenger ascorbate may also exert their action by decreasing the amount of O_2^- and thereby increasing the amount of NO. Thus, the 'redox-regulation' may simply be an in vitro artifact based on the extreme NO sensitivity of sGC and the O_2^--mediated inactivation of the dissolved atmospheric NO.

4 Modulators

4.1 ODQ

Recently, several modulators of sGC have been found. Garthwaite et al. (1995) showed that the quinoxalin derivative 1H-[1,2,4]oxadiazolo[4,3-a]-quinoxalin-1-one (ODQ) is a potent and selective inhibitor of sGC in brain slices. ODQ did not inhibit other cyclic nucleotide-forming enzymes as

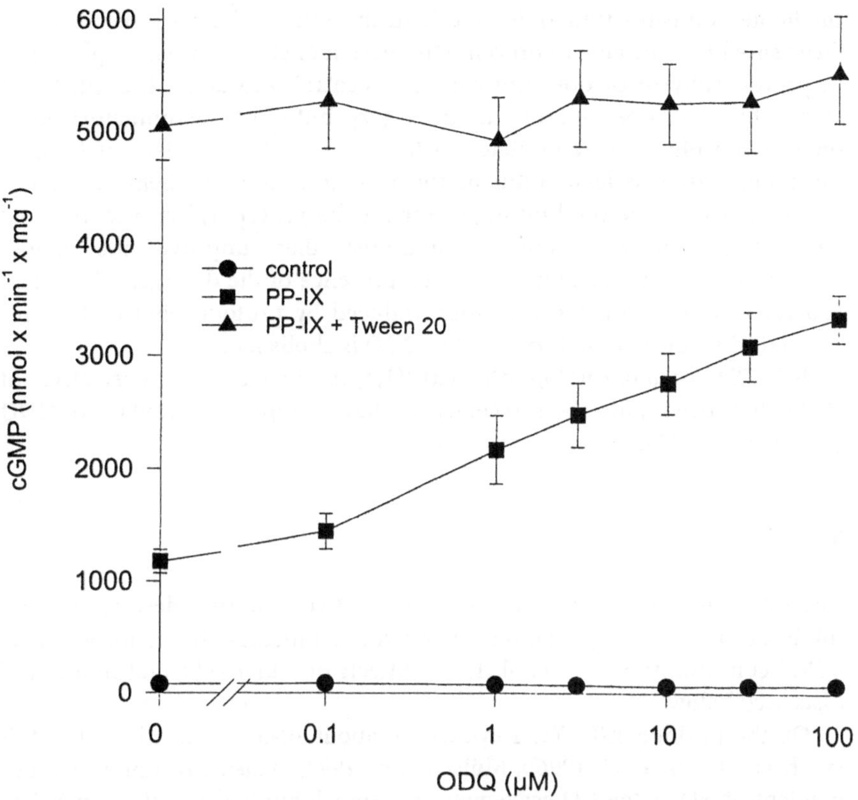

Fig. 5. Effect of ODQ on the stimulation of sGC by protoporphyrin IX. The non-stimulated enzyme activity is not altered by ODQ (circles). The stimulation induced by protoporphyrin IX is not inhibited but enhanced by ODQ (squares) whereas in the presence of the detergent Tween 20, which facilitates the heme-protoporphyrin IX exchange, ODQ does not influence sGC activity (triangles)

catalytic activity of membrane-bound guanylyl cyclases or adenylyl cyclases remained unchanged. More importantly, ODQ did not interfere with the stimulation of NO synthase and therefore represents an important tool to discriminate between cGMP-dependent and cGMP-independent NO signaling. Subsequently, inhibitory effects of ODQ on sGC have been demonstrated in a variety of other cells and tissues (Garthwaite et al. 1995; Brunner et al. 1996; Abiberges et al. 1997). Studies with purified sGC revealed that the drug binds in an NO-competitive manner and leads to an apparently irreversible inhibition of the stimulated enzyme, leaving basal activity almost unchanged. Spectral analysis suggests that the inhibitory effect of ODQ is due to oxidation of the heme iron (Schrammel et al. 1996). If the oxidation of the heme iron is the underlying mechanism of the inhibitory effect of ODQ, ODQ should not be able to prevent stimulation of sGC by protoporphyrin IX as protoporphyrin IX does not contain a central iron atom. Indeed, Fig. 5 shows that ODQ does not have an inhibitory, but rather a stimulatory effect on protoporphyrin IX-stimulated sGC. The stimulatory effect of ODQ is most likely due to facilitation of the protoporphyrin IX-heme exchange which appears to be the limiting factor for the protoporphyrin IX-induced stimulation. Likewise, under conditions that improve the heme-protoporphyrin IX exchange, i.e. in the presence of the detergent Tween 20 (Foerster et al. 1996), the activation induced by protoporphyrin IX is increased while the stimulatory effect of ODQ is abolished.

NS 2028 (oxadiazolo(3,4-d)benz(b)(1,4)oxazin-1-one), a derivative of ODQ, has been published recently to have properties similar to ODQ (Olesen et al. 1998).

4.2 YC-1

The substance YC-1, a benzylindazol derivative, was first described as an inhibitor of platelet aggregation which led to an increase in the intracellular cGMP concentration (Wu et al. 1995). Effects on other cells and tissues are discussed below.

On the purified sGC, YC-1 evoked an about 10-fold activation of cGMP synthesis (Friebe et al. 1996b; Mülsch et al. 1997). This activation was independent of NO as the NO scavenger oxyhemoglobin had no effect on YC-1-stimulated sGC. Despite the NO independence, YC-1 stimulation required the presence of the prosthetic heme group as YC-1 did not stimulate the heme-free enzyme. YC-1 exerted dramatic effects on the NO-stimulated enzyme as it increased the maximal NO-induced catalytic rate of sGC by about 40% and potentiated the stimulatory effect of submaximally activating NO concentrations. In the presence of YC-1, the concentration-response curve of NO was shifted to the left, indicating that YC-1 sensitized sGC towards NO. YC-1 did not only affect the NO-stimulated enzyme but also potentiated the stimulation of sGC induced by the NO-independent activator protoporphyrin IX. Moreover, even the poor activator CO was turned into an effective stimulator by YC-1 activating sGC to an extent similar to NO (Friebe et al. 1996b).

Although it was tempting to speculate about a direct YC-1-heme interaction, binding of YC-1 to the heme group is unlikely as YC-1 did not change the Soret absorption of basal or stimulated sGC (Friebe and Koesling, 1998). Taking advantage of the relatively slow dissociation rate, YC-1 was found to bind not only to the non-stimulated but also to the heme-depleted enzyme, suggesting an allosteric site on the enzyme. Obviously, although being able to bind to the heme-free enzyme, YC-1 only influenced the activated conformation of sGC and the YC-1 effect showed an absolute requirement for enzyme-bound porphyrin. The YC-1 effect on non-stimulated sGC may be explained by a small population of molecules in the activated state even in the absence of added activators. YC-1 slowed down the dissociation rates for NO and CO from the activated enzyme. Thus, YC-1 sensitized the enzyme towards its gaseous activators by decreasing the dissociation rate of the heme ligands (Friebe and Koesling, 1998).

Taken together, the effects of YC-1 on sGC suggest the existence of a so far unidentified allosteric site on the enzyme which modulates the catalytic rate and the responsiveness towards heme ligands. Ligands to this allosteric site may represent a novel class of drugs which may exert vasodilatory, anti-aggregatory and possibly other yet unknown effects by sensitizing sGC towards its physiological activator NO. In contrast to the NO donors commonly used in the treatment of coronary heart disease, YC-1-like substances would exert their action mainly at the sites of endogenous NO production. Moreover, one could speculate about the existence of a naturally occurring allosteric modulator of sGC which may alter the sensitivity of the enzyme towards NO. In case of endogenously generated CO concentrations reaching relevant levels, sGC may be significantly stimulated by CO in the presence of such an endogenous modulator.

YC-1 effects have now been demonstrated in a variety of cells and tissues. In vascular smooth muscle cells, YC-1 has been shown to increase cGMP and to induce a concentration-dependent relaxation of endothelial-free rat aortic rings precontracted by phenylephrine (Mülsch et al. 1997; Wegener and Nawrath, 1997). In addition, YC-1 has been shown to inhibit acetylcholine release and excitatory motor transmission in the guinea pig ileum (Hebeiss and Kilbinger, 1998). We have studied the possible synergistic effect of YC-1 and NO in intact platelets (Friebe et al. 1998). YC-1 together with NO or CO led to complete inhibition of platelet aggregation at concentrations that were ineffective by themselves (Fig. 6). The synergistic effect of YC-1 and NO was underlined by a drastic increase in intraplatelet cGMP. Whereas maximally effective NO or YC-1 concentrations only led to about 13-fold increases in cGMP, both drugs together resulted in an over 1300-fold rise of the cGMP content. Similar results were obtained using CO instead of NO.

In addition to the sGC stimulating effect, YC-1 led to an inhibition of cGMP-hydrolyzing phosphodiesterases in platelet cytosol. Thus, YC-1 obviously exerts a stimulatory effect on the cGMP-forming enzyme (sGC) and an inhibitory effect on cGMP-degrading enzymes (phosphodiesterases), there

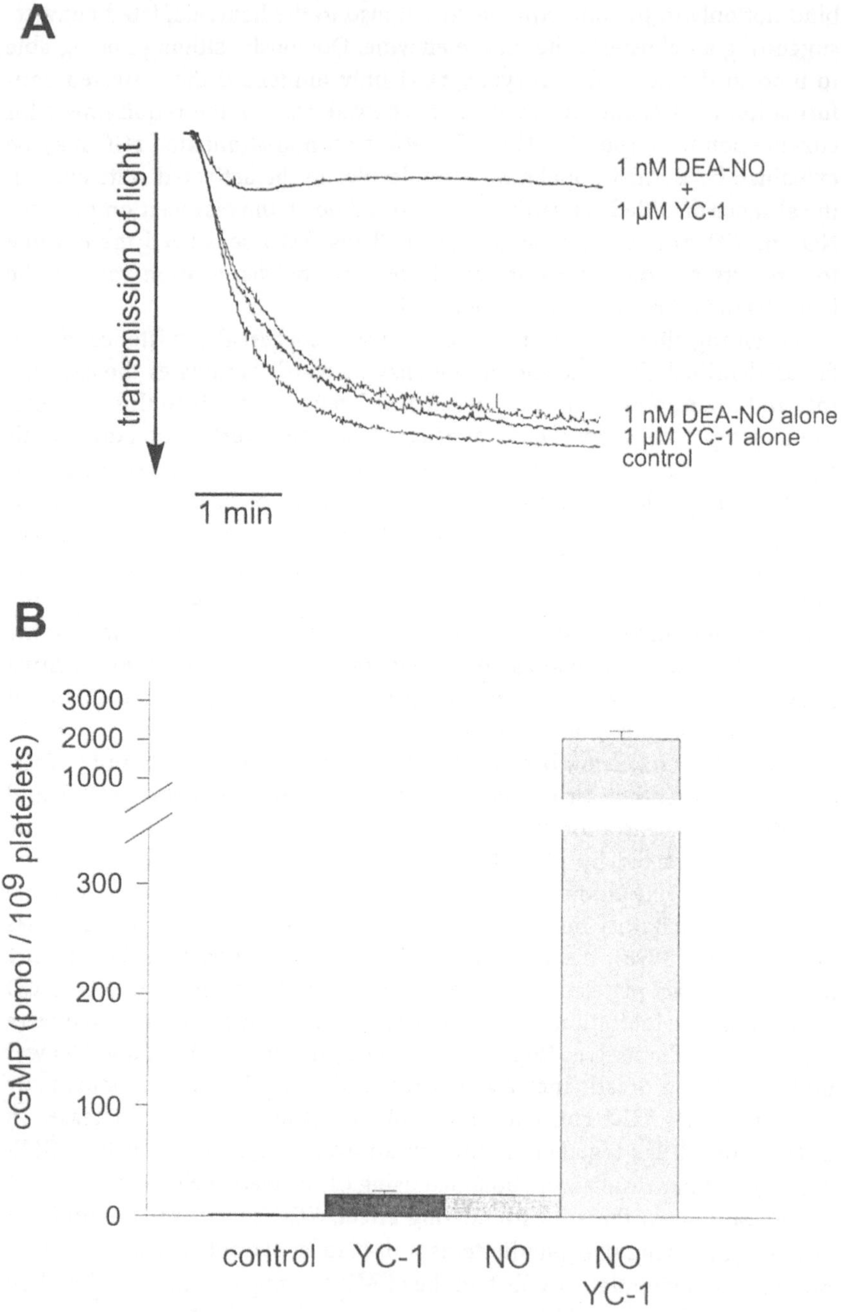
A
transmission of light
1 nM DEA-NO
+
1 μM YC-1
1 nM DEA-NO alone
1 μM YC-1 alone
control
1 min
B
cGMP (pmol / 10^9 platelets)
3000
2000
1000
300
200
100
0
control
YC-1
NO
NO
YC-1

by explaining the exceptional increase in the cGMP concentration in platelets. In that respect, YC-1 and YC-1-like substances are most interesting candidates as modulators of cGMP-mediated effects within the cardiovascular system and, indeed, YC-1 has been shown to inhibit platelet-rich thrombosis *in vivo* (Teng et al. 1997).

5 Concluding Remarks

Research on sGC in the last years has brought forth new and solid information about the enzyme. In contrast to the still increasing collection of membrane-bound guanylyl cyclases, only four subunits, two α (α_1, α_2) and two β (β_1, β_2) subunits of sGC have been identified. On the protein level, the physiological occurrence of two heterodimeric isoforms of sGC ($\alpha_1\beta_1$, $\alpha_2\beta_1$) has been demonstrated. The knowledge of the primary structure of sGC subunits revealed the close structural relationship to the membrane-bound guanylyl cyclases and adenylyl cyclases, the crystal structure of the catalytic adenylyl cyclase domain supplying a rough idea of the catalytic center of sGC.

By the analysis of purified enzyme ($\alpha_1\beta_1$) much has been learnt about the prosthetic heme group acting as the acceptor site for the enzyme's stimulator NO and the heme-dependent activation mechanism. The results with the new substance YC-1 revealed a novel mechanism to sensitize sGC towards NO which may have broad pharmacological or even physiological implications. Further studies will show whether the sensitivity of sGC towards NO is modulated in vivo by endogenous ligands at the YC-1 site. Adjusting NO sensitivity of sGC may represent a novel mechanism to modulate the cGMP response towards a given NO signal.

Fig. 6. Synergistic effect of YC-1 and NO on inhibition of platelet aggregation and cGMP levels. **A:** Platelet aggregation induced by 1 μM U 46619 was neither altered by 1 nM diethylamine-NO (DEA-NO) nor by 1 μM YC-1 whereas the combination of both led to a complete inhibition of aggregation. **B:** Cyclic GMP in platelets was only increased about 13-fold by maximally effective DEA-NO or YC-1 concentrations; when both components were applied together the cGMP levels in intact platelets were increased over 1300-fold

References

Abiberges N, Hovemadsen L, Fischmeister R, Mery PF (1997) A comparative study of the effects of three guanylyl cyclase inhibitors on the L-type Ca^{2+} and muscarinic K^{+} currents in frog. Br J Pharmacol 121:1369–1377

Arnold WP, Mittal CK, Murad F (1977) Nitric oxide activates guanylate cyclase and increases guanosine 3':5'-cyclic monophosphate levels in various tissue preparations. Proc Natl Acad Sci USA 74:3203–3207

Ashman DF, Lipton R, Melicow MM, Price TD (1963) Isolation of adenosine 3':5'-monophosphate and guanosine 3':5'-monophosphate from rat urine. Biochem Biophys Res Commun 11:330–334

Behrends S, Harteneck C, Schultz G, Koesling D (1995) A variant of the α_2 subunit of soluble guanylyl cyclase contains an insert homologous to a region within adenylyl cyclases and functions as a dominant negative protein. J Biol Chem 270:21109–21113

Böhme E, Graf H, Schultz G (1978) Effects of sodium nitroprusside and other smooth muscle relaxants on cyclic cGMP formation in smooth muscle and platelets. Adv Cyclic Nucleotide Res 9:131–143

Böhme E, Gerzer R, Grossmann G, Herz J, Mülsch A, Spies C, Schultz G (1983) Regulation of soluble guanylyl cyclase activity. In: Hormones and Cell Regulation Vol. 7 (Dumont JE, Nunez J, Denton RM, eds) Elsevier Biomedical Press, 147–161

Böhme E, Grossmann G, Herz J, Mulsch A, Spies C, Schultz G (1984) Regulation of cyclic GMP formation by soluble guanylate cyclase: stimulation by NO-containing compounds. Adv Cyclic Nucleotide Protein Phosphorylation Res 17:259–266

Brandwein HJ, Lewicki JA, Murad F (1981) Reversible inactivation of guanylate cyclase by mixed disulfide formation. J Biol Chem 256:2958–2962

Braun T, Dods RF (1975) Development of a Mn-2+-sensitive, "soluble" adenylate cyclase in rat testis. Proc Natl Acad Sci USA 72:1097–1101

Brüne B, Ullrich V (1987) Inhibition of platelet aggregation by carbon monoxide is mediated by activation of guanylate cyclase. Mol Pharmacol 32:497–504

Brüne B, Schmidt K-U, Ullrich V (1990) Activation of soluble guanylate cyclase by carbon monoxide and inhibition by superoxide anion. Eur J Biochem 192:683–686

Brunner F, Schmidt K, Nielsen EB, Mayer B (1996) Novel guanylyl cyclase inhibitor potently inhibits cyclic GMP accumulation in endothelial cells and relaxation of bovine pulmonary artery. J Pharmacol Exp Ther 277:48–53

Buechler WA, Nakane M, Murad F (1991) Expression of soluble guanylate cyclase activity requires both enzyme subunits. Biochem Biophys Res Commun 174:351–357

Burstyn JN, Yu AE, Dierks EA, Hawkins BK, Dawson JH (1995) Studies of the heme coordination and ligand binding properties of soluble guanylyl cyclase (sGC): Characterization of Fe(II)sGC and Fe(II)sGC(CO) by electronic absorption and magnetic circular dichroism spectroscopies and failure of CO to activate the enzyme. Biochemistry 34:5906–5903

Chinkers M, Garbers DL, Chang M-S, Lowe DG, Chin H, Goeddel DV, Schulz S (1989) A membrane form of guanylate cyclase is an atrial natriuretic peptide receptor. Nature 338:78–83

Craven PA, DeRubertis FR (1978) Restoration of the responsiveness of purified guanylate cyclase to nitrosoguanidine, nitric oxide, and related activators by heme and hemeproteins. Evidence for involvement of the paramagnetic nitrosyl-heme complex in enzyme activation. J Biol Chem 253:8433–8443

DeRubertis FR, Craven PA (1977) Activation of hepatic guanylate cyclase by N-methyl-N'-nitro-N-nitrosoguanidine. Effects of thiols, N-ethylmaleimide, and divalent cations. J Biol Chem 252:5804–5814

Dessauer CW, Gilman AG (1996) Purification and characterisation of a soluble form of mammalian adenylyl cyclase. J Biol Chem 271:16967–16974

Dierks EA, Burstyn JN (1996) Nitric oxide (NO), the only nitrogen monoxide redox form capable of activating soluble guanylyl cyclase. Biochem Pharmacol 51:1593–1600

Foerster J, Harteneck C, Malkewitz J, Schultz G, Koesling D (1996) A functional heme-binding site of soluble guanylyl cyclase requires intact N-termini of a_1 and b_1 subunits. Eur J Biochem 240:380–386

Friebe A, Malkewitz J, Schultz G, Koesling D (1996a) Positive side-effects of pollution? Nature 382:120

Friebe A, Schultz G, Koesling D (1996b) Sensitizing soluble guanylyl cyclase to become a highly CO-sensitive enzyme. EMBO J 15:6863–6868

Friebe A, Wedel B, Foerster J, Harteneck C, Malkewitz J, Schultz G, Koesling D (1997) Function of conserved cysteine residues on soluble guanylyl cyclase. Biochemistry 36:1194–1198

Friebe A, Koesling D (1998) Mechanism of YC-1-induced activation of soluble guanylyl cyclase. Mol Pharmacol 53:123–127

Friebe A, Koesling D (1998) Stimulation of soluble guanylyl cyclase by superoxide dismutase is mediated by NO. Biochem 335:527–531

Friebe A, Müllershausen F, Smolenski A, Walter U, Schultz G, Koesling D (1998) YC-1 potentiates NO- and CO-induced cGMP effects in human platelets. Mol Pharmacol 54

Garthwaite J, Charles SL, Chess-Williams R (1988) Endothelium-derived relaxing factor release on activation of NMDA receptors suggests role as intercellular messenger in the brain. Nature 336:385–388

Garthwaite J, Southam E, Boulton CL, Nielsen EB, Schmidt K, Mayer B (1995) Potent and selective inhibition of nitric oxide-sensitive guanylyl cyclase by 1H-[1,2,4]oxadiazolo[4,3-a]quinoxalin-1-one (ODQ). Mol Pharm 48:185–188

Gerzer R, Böhme E, Hofmann F, Schultz G (1981) Soluble guanylate cyclase purified from bovine lung contains heme and copper. FEBS Letters 132:71–74

Gerzer R, Hofmann F, Schultz G (1981) Purification of a soluble, sodium-nitroprussid-stimulated guanylate cyclase from bovine lung. Eur J Biochem 116:479–486

Giuili G, Scholl U, Bulle F, Guellaen (1992) Molecular cloning of the cDNAs coding for the two subunits of soluble guanylyl cyclase from human brain. FEBS Letters 304:83–88

Goldberg ND, O'Dea RF, Haddox MK (1973) Cyclic GMP. Adv Cycl Nucl Res 3:155–223

Gupta G, Azam M, Yang L, Danziger RS (1997) The β_2 subunit inhibits stimulation of the α_1/β_1 form of soluble guanylyl cyclase by nitric oxide. Potential relevance to regulation of blood pressure. J Clin Invest 100:1488–1492

Hardman JG, Sutherland EW (1969) Guanyl cyclase, an enzyme catalyzing the formation of guanosine 3',5'-monophosphate from guanosine trihosphate. J Biol Chem 244:6363–6370

Harteneck C, Koesling D, Söling A, Schultz G, Böhme E (1990) Expression of soluble guanylate cyclase: catalytic activity requires two subunits. FEBS Lett 272:221–223
Harteneck C, Wedel B, Koesling D, Malkewitz J, Böhme E, Schultz G (1991) Molecular cloning and expression of a new a-subunit of soluble guanylyl cyclase. FEBS Lett 292:217–222
Hebeiss K, Kilbinger H (1998) Nitric oxide-sensitive guanylyl cyclase inhibits acetylcholine release and excitatory motor transmission in guinea-pig ileum. Neuroscience 82:623–629
Humbert P, Niroomand F, Fischer G, Mayer B, Koesling D, Hinsch KD, Gausepohl H, Frank R, Schultz G, Böhme E (1990) Purification of soluble guanylyl cyclase from bovine lung by a new immunoaffinity chromatographic method. Eur J Biochem 190:273–278
Ignarro LJ, Degnan JN, Baricos WH, Kadowitz PJ, Wolin MS (1982) Activation of purified guanylate cyclase by nitric oxide requires heme. Comparison of heme-deficient, heme-reconstituted and heme-containing forms of soluble enzyme from bovine lung. Biochim Biophys Acta 718:49–59
Ignarro LJ, Wood KS, Wolin MS (1982) Activation of purified soluble guanylate cyclase by protoporphyrin IX. Proc Natl Acad Sci USA 79:2870–2873
Ignarro LJ, Adams JB, Horwitz PM, Wood KS (1986) Activation of soluble guanylate cyclase by NO-hemoproteins involves NO-heme exchange: comparison of heme-containing and heme-deficient enzyme forms. J Biol Chem 261:4997–5002
Ignarro, LJ, Buga GM, Wood KS, Byrns RE, Chandhuri G (1987): Endothelium-derived relaxing factor produced and released from artery and vein is nitric oxide. Proc Natl Acad Sci USA 84: 9265–9269
Ishikawa E, Ishikawa S, Davis JW, Sutherland EW (1969) Determination of guanosine 3',5'-monophosphate in tissues and of guanyl cyclase in rat intestine. J Biol Chem 244:6371–6376
Kharitonov VG, Russwurm M, Magde D, Sharma VS, Koesling D (1997) Dissociation of nitric oxide from soluble guanylyl cyclase. Biochem Biophys Res Commun 239:284–286
Koesling D, Böhme E, Schultz G (1991) Guanylyl cyclases, a growing family of signal transducing enzymes. FASEB J 5:2785–2791
Lincoln TM (1989) Cyclic GMP and mechanisms of vasodilation. Pharmacol Ther 41:479–502
Liu Y, Ruoho AE, Rao VD, Hurley JH (1997) Catalytic mechanism of the adenylyl and guanylyl cyclase: Modeling and mutational analysis. Proc Natl Acad Sci USA 94:13414–13419
Maines MD (1988) Heme oxygenase: function, multiplicity, regulatory mechanism, and clinical applications. FASEB J 2:2257–2568
Maines MD (1997) The heme oxygenase system: a regulator of second messenger gases. Annu Rev Pharmacol Toxicol 37: 517–554
Marks GS, Brien JF, Nakatsu KB and McLaughlin E (1991) Does carbon monoxide have a physiological function? Trends Pharmacol Sci 12:185–188
Mikami T, Kusakabe T, Suzuki N (1998) Molecular cloning of cDNAs and expression of mRNAs encoding alpha and beta subunits of soluble guanylyl cyclase from medaka fish Oryzias latipes. Eur J Biochem 253:42–48
Mittal CK, Murad F (1977) Formation of adenosine 3':5'-monophosphate by preparations of guanylate cyclase from rat liver and other tissues. J Biol Chem 252:3136–3140

Moncada S, Higgs EA (1995) Molecular mechanisms and therapeutic strategies related to nitric oxide. FASEB J 9:1319–1330
Mülsch A, Bauersachs J, Schäfer A, Stasch JP, Kast R, Busse R (1997) Effect of YC-1, an NO-independent, superoxide-sensitive stimulator of soluble guanylyl cyclase, on smooth muscle responsiveness to nitrovasodilators. Br J Pharmacology 120:681–689
Nakane M, Arai K, Saheki S, Kuno T, Buechler W, Murad F (1990) Molecular cloning and expression of cDNAs coding for soluble guanylyl cyclase from rat lung. J Biol Chem 265:16841–16845
O'Dell TJ, Hawkins RD, Kandel ER, Arancio O (1991) Tests of the roles of two diffusible substances in long-term potentiation: evidence for nitric oxide as a possible early retrograde messenger. Proc Natl Acad Sci USA 88:11285–11289
Ohlstein EH, Wood KS, Ignarro LJ (1982) Purification and properties of heme-deficient hepatic soluble guanylate cyclase: effects of heme and other factors on enzyme activation by NO, NO-heme, and protoporphyrin IX. Arch Biochem Biophys 218:187–198
Olesen SP, Drejer J, Axelsson O, Moldt P, Bang L, Nielsen-Kudsk JE, Busse R, Mülsch A (1998) Characterization of NS 2028 as a specific inhibitor of soluble guanylyl cyclase. Br J Pharmacol 123:299–309
Palmer RMJ, Ferrige AG, Moncada S (1987) Nitric oxide release accounts for the biological activity of endothelium-derived relaxing factor. Nature 327:524–526
Prabhakar S, Short DB, Scholz NL, Goy MF (1997) Identification of nitric oxide-sensitive and -insensitive forms of cytoplasmatic guanylate cyclase. J Neurochem 69:1650–1660
Russwurm M, Behrends S, Harteneck C, Koesling D (1998) Functional properties of a naturally occurring isoform of soluble guanylyl cyclase. Biochem 335:125–130
Schmidt H, Walter U (1994) NO at work. Cell 78:919–925
Schrammel A, Behrends S, Schmidt K, Koesling D, Mayer B (1996) Characterisation of 1H-[1,2,4]oxadiazolo[4,3-a]quinoxalin-1-one (ODQ) as a heme site inhibitor of nitric oxide-sensitive guanylyl cyclase. Mol Pharmacol 50:1–5
Schultz G, Böhme E, Munske K (1969) Guanyl cyclase. Determination of enzyme activity. Life Sci 8:1323–1332
Shah S, Hyde DR (1995) Two Drosophila genes that encode the a and b subunits of the brain soluble guanylyl cyclase. J Biol Chem 270:15368–15376
Stone JR, Marletta MA (1994) Soluble guanylyl cyclase from bovine lung: activation with nitric oxide and carbon monoxide and spectral characterisation of the ferrous and ferric states. Biochemistry 33:5636–5640
Stone JR, Marletta MA (1995) Heme stoichiometry of heterodimeric soluble guanylate cyclase. Biochemistry 34:14668–14674
Stone JR, Marletta MA (1998) Synergistic activation of soluble guanylate cyclase by YC-1 and carbon monoxide: implications for the role of cleavage of the iron-histidine bond during activation by nitric oxide. Chem Biol 5:255–261
Sunahara RK, Dessauer CW, Gilman AG (1996) Complexity and diversity of mammalian adenylyl cyclases. Annu Rev Pharmacol Toxicol 36:461–480
Sunahara RK, Beuve A, Tesmer JJG, Sprang SR, Garbers DL, Gilman AG (1998) Exchange of substrate and inhibitor specificities between adenylyl and guanylyl cyclases. J Biol Chem 273:16332–16338
Tang WJ, Stanzel M, Gilman AG (1995) Truncation and alanine-scanning mutants of type I adenylyl cyclase. Biochemistry 34:14563–14572

Tawata M, Aida K, Noguchi T, Ozaki Y, Kume S, Sasaki H, Chin M, Onaya T (1992) Anti-platelet action of isoliquiritigenin, an aldose reductase inhibitor in licorice. Eur J Pharmacol 212: 87–92

Teng CM, Wu CC, Ko FN, Lee FY, Kuo SC (1997) YC-1, a nitric oxide-independent activator of soluble guanylate cyclase, inhibits platelet-rich thrombosis in mice. Eur J Pharmacol 320:161–166

Tesmer JJG, Sunahara RK, Gilman AG, Sprang SR (1997) Crystal structure of the catalytic domains of adenylyl cyclase in a complex with $G_{s\alpha} \cdot GTP\gamma S$. Science 278:1907–1916

Thompson DK, Garbers DL (1995) Dominant negative mutations of the guanylyl cyclase-A receptor. J Biol Chem 270:425–430

Tomita T, Tsuyama S, Imai Y, Kitagawa T (1997) Purification of bovine soluble guanylate cyclase and ADP-ribosylation on its small subunit by bacterial toxins. J Biochemistry 122:531–536

Tucker CL, Hurley JH, Miller TR, Hurley JB (1998) Two amino acid substitutions convert a guanylyl cyclase, RetGC-1, into an adenylyl cyclase. Proc Natl Acad Sci USA 95:5993–5997

Waldman SA, Murad F (1987) Cyclic GMP synthesis and function. Pharmacol Rev 39:163–196

Walter U (1989) Physiological role of cGMP and cGMP-dependent protein kinase in the cardiovascular system. Rev Physiol Biochem Pharmacol 113:42–88

Wedel B, Humbert P, Harteneck C, Foerster J, Malkewitz J, Böhme E, Schultz G, Koesling D (1994) Mutation of His-105 of the β_1 subunit yields a nitric oxide-insensitive form of soluble guanylyl cyclase. Proc Natl Acad Sci USA 91:2592–2596

Wedel B, Harteneck C, Foerster J, Friebe A, Schultz G, Koesling D (1995) Functional domains of soluble guanylyl cyclase. J Biol Chem 270:24871–24875

Wegener JW, Nawrath H (1997) Differentiell effects of isoliquiritigenin and YC-1 in rat aortic smooth muscle. Eur J Pharmacol 3203:89–91

White AA, Aurbach GD (1969) Detection of guanyl cyclase in mammalian tissues. Biochim Biophys Acta 191:686–697

Wilson EM, Chinkers M (1995) Identification of sequences mediating guanylyl cyclase dimerisation. Biochemistry 34:4696–4701

Wu CC, Ko FN, Kuo SC, Lee FY, Teng CM (1995) YC-1 inhibited human platelet aggregation through NO-independent activation of soluble guanylate cyclase. Br J Pharmacol 116:1973–1978

Yoshikawa S, Miyamoto I, Aruga J, Furuichi T, Okano H, Mikoshiba K (1993) Isolation of a Drosophila gene encoding a head-specific guanylyl cyclase. J Neurochem 60:1570–1573

Yu S, Avery L, Baude E, Garbers DL (1997) Guanylyl cyclase expression in specific sensory neurons: a new family of chemosensory receptors. Proc Natl Acad Sci USA 94:3384–3387

Yuen PST, Potter LR, Garbers DL (1990) A new form of guanylyl cyclase is preferentially expressed in rat kidney. Biochemistry 29:10872–10878

Zhao Y, Marletta MA (1997) Localization of the heme binding region in soluble guanylate cyclase. Biochemistry 36:15959–15964

Zhao Y, Schelvis JP, Babcock GT, Marletta MA (1998) Identification of histidine 105 in the β_1 subunit of soluble guanylate cyclase as the heme proximal ligand. Biochemistry 37:4502–4509

Zhuo M, Hawkins RD (1995) Long-term depression: a learning-related type of synaptic plasticity in the mammalian central nervous system. Rev Neurosci 6:259–277

Cyclic GMP as Substrate and Regulator of Cyclic Nucleotide Phosphodiesterases (PDEs)

D. M. Juilfs[1], S. Soderling[2], F. Burns[2], and J. A. Beavo[2]

[1] Parke-Davis Pharmaceutical Research, 2800 Plymouth Road, Ann Arbor, MI 48105
[2] Department of Pharmacology, University of Washington, Seattle, WA 98195

Contents

1 Introduction

The regulation of cyclic nucleotide phosphodiesterases (PDEs), those enzymes responsible for the degradation of the second messengers, cAMP and cGMP, is important in mediating the diverse effects of various hormones, neurotransmitters and other agonists, including light. The intensity and duration of the cellular response to these effectors is dependent upon the activity of the various PDE isoforms expressed within a cell. During the past two decades extensive efforts have been made towards identifying and characterizing these PDEs. The description and the characterization of the intracellular signaling pathways that PDEs regulate has led to the growing realization of the importance they play in regulating physiologic responses. To date eight families of eukaryotic PDEs have been identified and characterized. An updated list of the currently accepted nomenclature for these PDEs can be found on the World Wide Web at the following URL: http://weber.u.washington.edu/~pde/. Six of the PDE families are capable of either being regulated by cGMP or hydrolyzing cGMP (Beavo, 1995). Many of these PDEs have been demonstrated to be important physiologic regulators of a number of pathways including vision (PDE6) (Farber, 1995), the male erectile response (PDE5) (Boolell et al. 1996), adrenal steroidogenesis (PDE2) (MacFarland et al. 1991), cardiac contractility (PDE2 and PDE3) (RivetBastide et al. 1997), and insulin / IGF-1 action (Zhao et al. 1997). Each of the PDE families is distinguished by a unique primary structure, regulatory properties, hydrolytic specificity and inhibitor profiles. Moreover, each family member exhibits higher sequence homology to its counterpart across species than to members of other families within a species. Most of the PDE families consist of proteins derived from more than one gene. In addition, multiple splice variants occur for many of the PDE genes resulting in proteins with unique termini. It is likely that more individual members and entire families are yet to be discovered.

All PDEs have an approximately 270 amino acid region of homology near the carboxy terminus believed to encompass the catalytic domain (Charbonneau, 1990). This region confers a unique substrate specificity for each PDE family. The cGMP-binding PDEs (PDE5), the photoreceptor PDEs (PDE6) and the recently described PDE9 family specifically hydrolyze cGMP. The PDE9 family is the newest member of the PDE superfamily and will be discussed in more detail in a later section. Unlike the cGMP-specific PDEs, the calcium/calmodulin-stimulated PDEs (PDE1) and the cGMP-stimulated PDEs (PDE2s) can hydrolyze both cyclic nucleotides. Although the catalytic domain of the cGMP-inhibited PDE (PDE3) can bind both cyclic nucleotides, it hydrolyzes only cAMP efficiently. The PDE4 and PDE7 families are specific for cAMP hydrolysis and will not be discussed further in this review.

In addition to differences in their substrate specificity, the hydrolytic activity of several PDE families is regulated in an allosteric manner by the interaction of cGMP to a region upstream of the catalytic domain. The cGMP-stimulated PDEs (PDE2), cGMP-binding PDEs (PDE5), (McAllister et al. 1993; Thomas et al. 1990a) and the photoreceptor PDEs (PDE6) all contain a non-catalytic cGMP binding domain that is distinct from the catalytic domain (Charbonneau et al. 1990). This region consists of two homologous repeats encoding cGMP binding domains that are similar between the three PDE proteins. While the activity of PDE2 is clearly stimulated in the presence of low concentrations of cGMP, the role of cGMP interaction with the binding domains in the regulation of PDE5 and PDE6 activity is not completely understood. The cGMP inhibitied PDEs, PDE3, are also regulated by cGMP. These PDEs, however, do not have a cGMP binding domain distinct from the catalytic domain. Cyclic GMP is thought to inhibit cAMP hydrolysis by competition at the catalytic domain, but is itself only slowly hydrolyzed. This review will focus primarily on the cGMP-stimulated isoforms of PDE (PDE2) with shorter discussions on PDE1, PDE3, PDE5, and PDE9 in relation to cGMP hydrolysis and regulation.

2 Cyclic GMP-Stimulated Phosphodiesterases (PDE2)

A cGMP stimulated increase in cAMP hydrolysis was first described in the extracts from several different tissues in the early 1970s (Beavo et al. 1971; Franks and Macmanus, 1971). The activity was shown to be differentially distributed between the supernatant and particulate fractions of these tissues. This protein was eventually purified and more recently cDNA clones have been identified. However, very little progress in the way of identifying physiological roles for this protein has been made until recently. We now know of at least three PDE2 isoforms and the use of a recently discovered PDE2 inhibitor, EHNA, has greatly enhanced the progress in identifying roles of PDE2 in regulation of cellular functions (Podzuweit et al. 1995).

PDE2 was purified to homogeneity from the supernatant fractions of bovine heart, adrenal, liver and platelets using a cGMP affinity resin as the critical final purification step (Martins et al. 1982; Miot et al. 1985; Yamamoto et al. 1983). In addition, PDE2 was purified from the particulate fractions of bovine and rabbit brain using detergents or trypsin treatment to liberate the enzymatic activity from the membranes (Murashima et al. 1990; Whalin et al. 1988). All of these purified and partially purified PDE2 preparations, except for the liver protein, demonstrate similar enzyme kinetics (reviewed by (Manganiello et al. 1990)). PDE2 migrates at approximately 102–105 kDa on SDS-PAGE. Examination by native gel electrophoresis indicates that PDE2 exists as a dimer (Martins et al. 1982). This protein hydrolyzes both cAMP (K_m = 30 µM) and cGMP (K_m = 10 µM) at a similar maximum velocity from 100–150 µmol · min^{-1} · mg^{-1} of protein and displays positive cooperative kinetics which is a feature unique to the PDE2 family. The cGMP-stimulated enzyme purified from rat liver soluble and particulate fractions, however, differed in size at 66 kDa and had a V_{max} of 4 µmol · min^{-1} · mg^{-1} (Pyne et al. 1986) suggesting that this cGMP-stimulated activity may be a unique isoform of PDE2 possibly originating from a different gene.

The distinguishing characteristic of the PDE2 protein is the ability of cGMP to stimulate cAMP hydrolysis up to 50-fold in pure preparations with a K_a of about 0.5 µM. The extent of stimulation varies between tissue homogenates depending on the contribution of other PDE isoforms present in that tissue. Higher concentrations of cGMP inhibit cAMP hydrolysis, presumably due to competition of cAMP at the catalytic domain. Cyclic GMP appears to increase the affinity of the catalytic domain for cAMP with little change in V_{max} and converts the enzyme to one that follows linear Michaelis-Menton kinetics, suggesting that PDE2 contains a low affinity catalytic domain that probably changes conformation to a higher affinity state upon cGMP binding to its allosteric domain (Manganiello et al. 1990). This hypothesis was supported by cGMP-binding studies that showed two noncatalytic cGMP-binding sites that can be separated from the catalytic activity by chymotryptic cleavage (Stroop and Beavo, 1991). In addition, it appears that cGMP-binding causes a conformational change since in the absence of cGMP, PDE2 becomes resistant to chymotryptic cleavage (Stroop and Beavo, 1991); (Moss et al. 1977).

2.1
PDE2 Isoforms

Three isoforms of the cGMP-stimulated phosphodiesterase (PDE2) have been identified to date (Fig. 1). Primary amino acid sequence was obtained from PDE2 protein purified from bovine heart (Trong et al. 1990). This sequence was completed and verified by the nucleotide sequence of a PDE2 cloned from a bovine adrenal cDNA library (Sonnenburg et al. 1991) and was designated PDE2A1. Examination of the amino acid sequence reveals a catalytic domain 25–33% homologous to other cyclic nucleotide phosphodiesterases and a cGMP binding domain similar to those present in the cGMP binding PDE (PDE5) and the photoreceptor PDEs (PDE6s). A second PDE2 isoform, PDE2A2, was cloned from rat (Yang et al. 1994) and mouse (Juilfs, unpublished) brain cDNA libraries. The third isoform identified, PDE2A3, was cloned from bovine and mouse brain cDNA libraries (Juilfs, unpublished) and was recently cloned from a human brain cDNA library (Rosman et al. 1997). The three PDE2 isoforms are identical except for the first 17–25

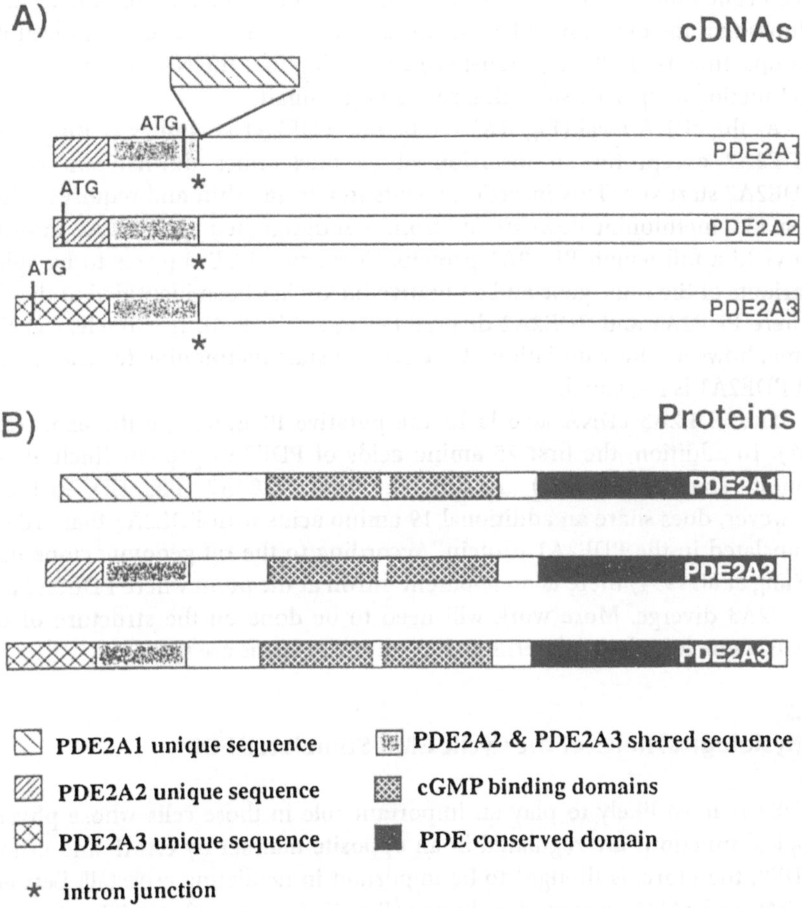

Fig. 1A, B. Schematic diagram of cDNA and protein sequences of PDE2 isoforms. **A** cDNA comparisons. PDE2A1 contains an insert that results in a unique start site for translation as compared to PDE2A2. PDE2A3 also lacks the PDE2A1 insert and has a unique amino terminus. These cDNAs appear to be splice variants of the same gene. **B** Protein comparisons. All three proteins have unique amino termini, however, PDE2A2 and PDE2A3 share more sequence with each other than with PDE2A1

amino acids. These differences are thought to be responsible for the soluble versus particulate nature of the PDE2 proteins. Since the majority of cGMP-stimulated activity in bovine heart and adrenal was found in the supernatant fractions (Martins et al. 1982), PDE2A1 has been called the 'soluble' isoform. In both the rabbit and bovine brain greater than 75% of the cGMP-stimulated activity was associated with particulate fractions (Murashima et al. 1990; Whalin et al. 1988). Therefore, both PDE2A2 and PDE2A3 have been called the 'membrane' or 'particulate' forms of PDE2. Preliminary data shows that the expressed PDE2 isoforms are targeted to different subcellular compartments (Juilfs, unpublished) suggesting that the N-termini are likely to function as specific subcellular targeting domains.

At the cDNA level (Fig. 1A) the bovine PDE2A1 is similar to the rodent PDE2A2 except for an insertion of 62 nucleotides downstream of the PDE2A2 start site. This insertion results in a frame shift and requires initiation at a methionine downstream from that designated for PDE2A2 in order to yield a full length PDE2A1 protein. These two PDEs appear to be splice variants of the same gene and a putative intron has been identified at the site where PDE2A1 and PDE2A2 diverge (Yang et al. 1994). It is unclear at this time, however, how initiation at the second start methionine for translation of PDE2A1 is regulated.

The PDE2A3 cDNA also lacks the putative PDE2A1 specific exon (Fig. 1A). In addition, the first 25 amino acids of PDE2A3 protein (including a unique start methionine) are different from PDE2A2 (Fig. 1B). PDE2A3, however, does share an additional 19 amino acids with PDE2A2 that are not translated in the PDE2A1 protein. According to the rat genomic clone data (Yang et al. 1994) there is no apparent intron at the point where PDE2A2 and PDE2A3 diverge. More work will need to be done on the structure of the PDE2 gene in order to determine how these isoforms are made.

2.2 Physiological Roles of the Cyclic GMP-Stimulated PDE

PDE2 is most likely to play an important role in those cells whose physiological functions are regulated in an opposite manner by cAMP and cGMP. PDE2, therefore, is thought to be important in mediating crosstalk between cGMP and cAMP regulated pathways (Fig. 2). For example, atrial natriuretic peptide (ANP), a peptide hormone that activates a membrane guanylyl cyclase receptor and increases cGMP levels, has been shown to reduce cAMP levels by the activation of PDE2 in bovine aortic vascular endothelial cells (Kishi et al. 1994) and PC12 cells (Whalin et al. 1990). Isoproterenol or PGE1 stimulated cAMP accumulation in human skin fibroblasts (Lee et al. 1988) is

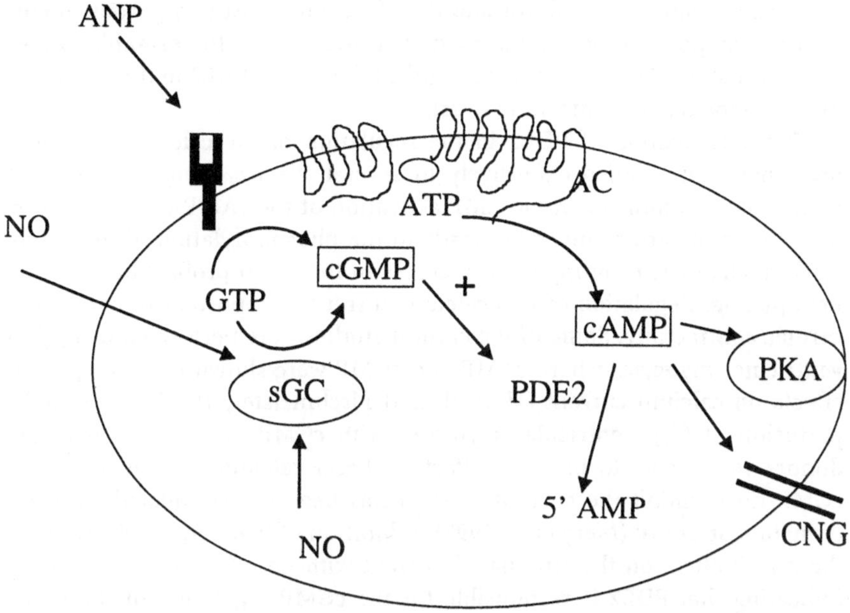

Fig. 2. Schematic diagram suggesting role for PDE2 in modulating cAMP mediated pathways. Many effectors stimulate cAMP synthesis by adenylyl cyclase (AC) via activation of the seven transmembrane G-protein coupled receptors. cAMP then stimulates downstream signaling by binding to protein kinase A (PKA) or cyclic nucleotide gated channels (CNG). Activation of guanylyl cyclases (GC) can occur by interaction of nitric oxide (NO) with the soluble isoforms or peptide hormones, like the atrial natriuretic peptide (ANP) with the receptor transmembrane isoforms. This increase in cGMP can stimulate PDE2 thereby increasing cAMP hydrolysis and down regulating cAMP signaling

also reduced by an ANP mediated increase in phosphodiesterase activity. In these examples the ANP mediated decrease in cAMP has not yet been associated with a specific physiological function, however, in cultured bovine adrenal glomerulosa cells an ANP mediated decrease in ACTH-stimulated cAMP levels results in a decrease in aldosterone secretion (MacFarland et al. 1991).

The recent development of a PDE2 specific inhibitor, EHNA (erythro-9-(2-hydroxy-3-nonyl)-adenine) (Podzuweit et al. 1995), has supplied the means for many laboratories to more easily investigate those cellular pathways that are modulated by PDE2. EHNA inhibits PDE2 in a concentration dependent manner with an IC50 ranging from 0.8–2 μM which is at least 50-fold less than for other PDE isoforms. However, this compound is also an adenosine deaminase inhibitor and therefore, one must be rigorous in controlling for possible adenosine mediated effects (see for example (Rivet-Bastide et al. 1997). This group has studied the role of PDE2 in the regulation of the L-type calcium current in heart.

The participation of PDE2 in the regulation of the calcium current in heart has been studied extensively in several species, including frog, rat, guinea pig and human. Cyclic AMP activation of the cAMP-dependent protein kinase in cardiac myocytes leads to the phosphorylation of the L-type calcium channels resulting in an increase in the mean probablility of channel opening. Regulation of this calcium current is important in controlling cardiac contractility. Some of the earliest studies were performed using frog ventricular myocytes where cAMP and cGMP were shown to have opposite effects on calcium current (Hartzell and Fischmeister, 1986). Intracellular perfusion of frog ventricular myocytes with cGMP or application of NO donors was shown to have no effect on basal calcium current. However, cGMP does inhibit the current when it has been pre-stimulated with isoprenaline or cAMP (Mery et al. 1995). Inhibition of PDE2 by EHNA reverses the cGMP effect on the stimulated current with no effect on basal current indicating that PDE2 is responsible for the cGMP regulation of the cAMP stimulated calcium current.

It has become clear as this pathway has been studied in different species that each species examined utilizes a slightly different complement of PDE activities to regulate the calcium current. It is also likely that different regions of the heart will utilize different PDEs in order to fulfill the unique functions of each region. In rat ventricular and atrial myocytes PDE2 does not appear to be important in the regulation of either the basal or the cAMP-stimulated calcium current. Intracellular perfusion with cGMP, however, was shown to inhibit isoprenaline or cAMP-stimulated current by activating the cGMP-dependent protein kinase (Mery et al. 1991).

In guinea pig ventricular myocytes intracellular application of cGMP enhances rather than inhibits both basal and isoprenaline stimulated calcium current. This effect is attributed to the cGMP-inhibited PDE (PDE3) suggesting PDE3 rather than PDE2 is important in regulation of the calcium current in guinea pig myocytes (Shirayama and Pappano, 1996). In frog myocytes inhibition of PDE3 by milrinone has no effect.

In human atrial myocytes PDE2 appears to regulate only the basal calcium current while PDE3 is the predominant cAMP hydrolyzing enzyme as determined by inhibition by milrinone (RivetBastide et al. 1997). Both PDE2 and PDE3 are needed to maintain basal cAMP levels since inhibition of either one or the other leads to large increases in calcium current. PDE3, however, appears dominant since concentrations of cGMP that should maximally inhibit PDE3 and fully stimulate PDE2 still results in a substantial increase in calcium current. Therefore, the regulation of basal calcium current in human atrial myocytes is dependent on a unique balance of cGMP, cAMP and the activity of PDE2 and PDE3.

Studies performed on the SA and AV node examine upstream regulation of cGMP as well as the subsequent signaling events leading to the regulation of the calcium current (Han et al. 1995; Han et al. 1997). Using rabbit sinoatrial node myocytes this group showed that muscarinic stimulation was able to reverse isoproterenol stimulated calcium current and that this reversal was blocked by NOS inhibitors and by agents that inhibit cGMP production. Nonhydrolyzable cAMP analogues and the PDE inhibitor, IBMX, both stimulate the calcium current and block the cholinergic effect suggesting that a PDE is involved in mediating the cGMP effect on the cAMP regulated calcium current. Milrinone has no effect indicating that PDE2 is likely to be responsible. Further investigations are warranted on the differences in PDE2 and PDE3 expression is different regions of the heart and the effect of this on heart function.

Recently PDE2, through the use of EHNA, has been implicated in several other physiological functions. EHNA reverses the pulmonary vasoconstriction that follows a hypoxic challenge in the perfused rat lung (Haynes et al. 1996). An EHNA inhibitable PDE2 was shown to be present in the smooth muscle of rat lung and was shown to be at least partly responsible for this vasodilatory response. This suggests PDE2 might be involved in the modulation of smooth muscle tone in the rat lung.

Activation of PDE2 in platelets has been shown to be at least partly responsible for regulating cAMP levels (Dickinson et al. 1997). Both cAMP and cGMP inhibit thrombin induced platelet aggregation. Although PDE3 is the major cAMP hydrolyzing enzyme (Maurice and Haslam, 1990) PDE2 appears to be necessary to restrict increases in cAMP that occurs due to PDE3 inhibition. When cAMP levels are high, PDE2 plays the major role in cAMP breakdown.

PDE2 is also expressed in mouse thymocytes (Michie et al. 1996). In these cells PDE2 activity is rapidly and transiently decreased upon ligation of the T-cell antigen receptor in response to phytohematogluttin but not to anti-CD3 or anti-TCR. This data shows that PDE2 can be specifically and differentially regulated in some cell types. Although no cAMP functional effect was examined, the authors suggest that cAMP-mediated apoptosis or thymocyte differentiation might be modulated in these cells by regulation of PDE2 activity.

Nitric oxide donors and ANP have been shown to block H_2O_2 mediated increases in the permeability of porcine pulmonary artery endothelial cells.

EHNA inhibition of PDE2 reduces this effect (Suttorp et al. 1996). However, PDE2 is thought to be regulating permeability through the hydrolysis of cGMP rather than cAMP in this system suggesting that PDE2 in these cells could be functioning in a negative feedback response to cGMP synthesis rather than in the mediation of a cAMP regulated pathway.

Many of the physiological functions that involve regulation of PDE2 are very complex. In many cases, inhibition of PDE2 without first increasing intracellular cAMP levels has little or no effect. Therefore, regulation of PDE2 activity by cGMP is probably most important when cAMP levels are high. In addition, many of the effects listed above involve modulation of cyclic nucleotide levels by more than one PDE family. For example, regulation of platelet aggregation involves both PDE2 and PDE3 as described and PDE5 is also an important regulator of cGMP levels in platelets (Chiu et al. 1997; Ito et al. 1996). In addition, there is variability in the expression of many PDE isoforms between species. This must be taken into account when extrapolating data from one species to the next. For example, in human atria PDE3 plays a major role in the regulation of myocardial contractility (Rivet-Bastide et al. 1997) while in the frog ventricle PDE2 appears to be the major cAMP hydrolyzing enzyme. Finally, PDE expression also varies during differentiation as well as throughout development (Michie et al. 1996; Rybalkin et al. 1997)

2.3 Expression of PDE2 in Olfactory Neurons

Other tissues have been shown to express PDE2 at relatively high concentrations including in the olfactory neuroepithelium (Juilfs et al. 1997) and many neurons found throughout the brain (Repaske et al. 1993) suggesting that PDE2 might also be involved in neuronal signaling.

Odorant information is encoded by a series of intracellular signal transduction events in olfactory neurons within the nasal epithelium thought to be mediated primarily by the second messenger cAMP (Brunet et al. 1996). We recently showed that a small subset of these olfactory neurons express the cGMP-stimulated PDE2 and the membrane guanylyl cyclase-D receptor (GC-D) suggesting that cGMP in these neurons also can have an important regulatory function in olfactory signaling (Juilfs et al. 1997). Both PDE2 and GC-D are expressed in the olfactory cilia of these neurons where odorant signaling is initiated (Fig. 3). Although the ligand for GC-D has not been described, it is anatomically positioned in a such a way as to be able to interact with ligands in the nasal cavity. PDE2 also appears to be expressed in the

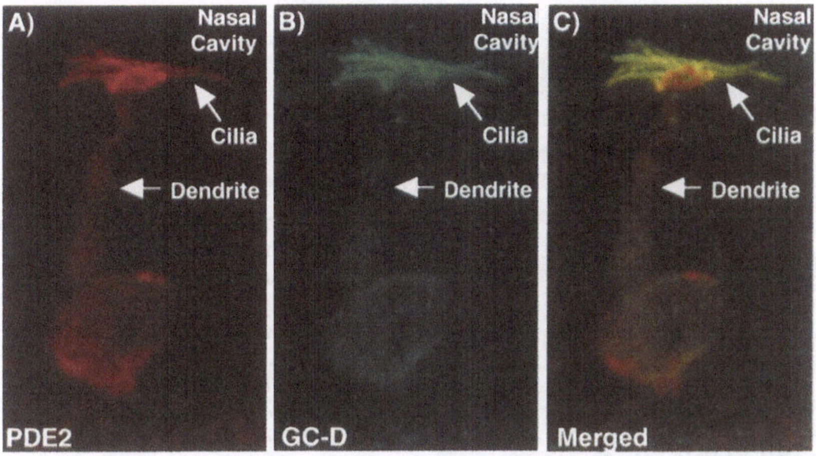

Fig. 3A–C. Expression PDE2 and GC-D in the same olfactory neurons. Olfactory tissue was double labeled with antisera developed against PDE2 (host-chicken) and GC-D (host-rabbit) and visualized with secondary antisera conjugated to rhodamine or fluorescein. The resulting immunofluorescence was visualized with a Biorad MRC-600 laser scanning confocal microscope. **A** PDE2 **B** GC-D **C** Merged image. PDE2 and GC-D are clearly observed in cilia originating from the same olfactory neuron upon merging of the individual labels

axons of the neurons suggesting that PDE2 may have a more diverse role in regulating olfactory signal transduction.

This subset of neurons is unique not only because they express these cGMP signal transduction proteins, but also because they lack the expression of several previously identified components of olfactory signal transduction cascades involving cAMP and calcium. These include the olfactory specific calcium/calmodulin-dependent PDE (PDE1C2) (Yan et al. 1995), an adenylyl cyclase III (Bakalyar and Reed, 1990), and a cAMP-specific PDE (PDE4A) (Cherry and Davis, 1995). Interestingly, we showed that these latter three proteins are expressed in the same neurons; however, their subcellular distribution is quite distinct (Fig. 4). PDE1C2 and adenylyl cyclase III are expressed almost exclusively in the olfactory cilia while PDE4A is present only in the cell bodies and axons. These data strongly suggest that selective compartmentalization of different phosphodiesterases and cyclases is an important feature for the regulation of signal transduction in most olfactory neurons.

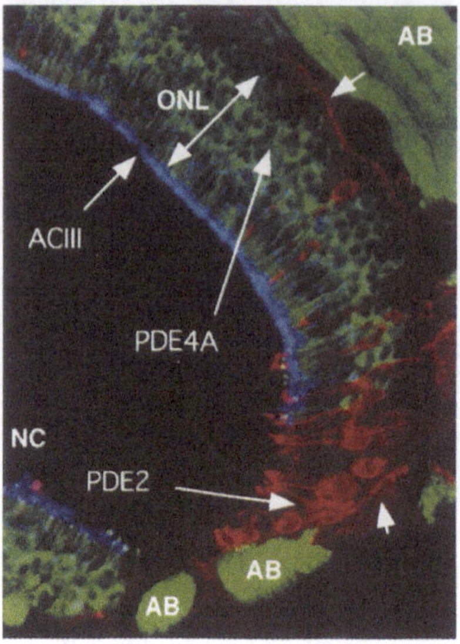

Fig. 4. PDE2 and GC-D are found in a subset of olfactory neurons distinct from those neurons expressing ACIII and PDE4A. Olfactory tissue was labeled simultaneously with antisera developed against PDE4A (*green*), PDE2 (*red*) and ACIII (*blue*). No neurons are present that express both PDE4A and PDE2. It has been shown that many neurons that express PDE4A also express ACIII and PDE1C2 in the cilia only (Juilfs, et al, 1997). ONL = olfactory neuronal layer, AB = axon bundle, NC = nasal cavity, arrow = PDE2 expressing axon fibers

It is estimated that there are greater than 500 different odorant receptor genes (Buck and Axel, 1991) expressed in specific olfactory neurons localized to broadly defined zones throughout the neuroepithelial layer. In addition, axons from olfactory neurons that express the same odorant receptor mRNA have been shown to project to single glomeruli in the olfactory bulb (Ressler et al. 1994; Vassar et al. 1994). This does not appear to be the case for those neurons expressing PDE2 and GC-D. Their axons project to a distinct group of glomeruli in the olfactory bulb that appear to be the same subset that have been termed 'necklace glomeruli' (Juilfs et al. 1997). A schematic showing the relative positions of the olfactory neurons and glomeruli expressing PDE2 in the nasal cavity is shown in Fig. 5. One of the necklace glomeruli, the modified glomerular complex, has been associated with suckling behavior in rat pups (Greer et al. 1982; Teicher et al. 1980). Other olfactory neurons and glomeruli defining unique subsets have been identified by immunohistochemistry with monoclonal antibodies designed against olfactory neuroepithelial specific proteins. Ring et al. have shown that a subset of neurons and glomeruli very similar to those that express PDE2 are labeled by the monoclonal antibody, MAb213 (Ring et al. 1997). They also showed that this subset of neurons are mature neurons, even though they are only weakly labeled by olfactory marker protein. These data suggest that the group of neurons expressing PDE2 may also express other proteins unique to this subset.

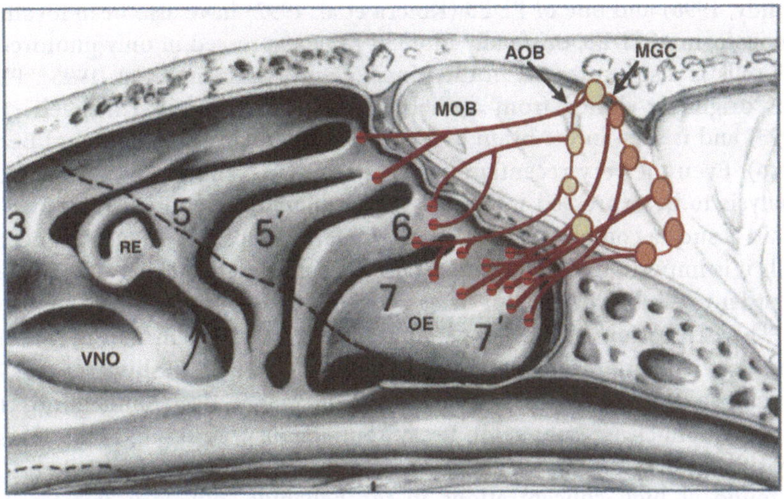

Fig. 5. Schematic drawing showing the arrangement of olfactory 'necklace' glomeruli that are innervated by axons expressing PDE2. Drawing of a sagittal section through the rat nasal cavity and anterior portion of the brain. Most PDE2 labeled neurons are positioned at the posterior portion of the olfactory epithelium (black dots). The PDE2 labeled axons from these neurons project to glomeruli in the posterior portion of the olfactory bulb to a group of glomeruli connected by fibers also labeled by PDE2 antisera. This group has been described as e the 'necklace' olfactory glomeruli. VNO = vomeronasal organ, RE = respiratory pithelium, OE = olfactory epithelium, MOB = main olfactory bulb, AOB = accessory olfactory bulb, MOB = main olfactory bulb, MGC = modified glomerular complex

The data implies that an olfactory signal transduction pathway specifically modulated by cGMP is present in a subset of neurons of the olfactory neuroepithelium. Furthermore, because the expression of signal transduction proteins and the organization of these neurons and glomeruli is different from those thought to be responsible for odorant detection these olfactory neurons may function in a role other than detection of specific odorants.

2.4 PDE2 Expression in Mouse Brain

The brain is one organ that appears to express at least one isoform from most if not all of the PDE families. Several isoforms of PDE1 (Yan et al. 1994; Yan et al. 1996) and PDE4 (Iona et al. 1998; Lobban et al. 1994; Mcphee et al. 1995) have been described in brain. Two isoforms of PDE3 (Reinhardt and

Bondy, 1996) and one of PDE5 (Kotera et al. 1997) have also been localized. An isoform of PDE6, originally thought to be expressed in only photoreceptor cells is also expressed in the pineal gland (Carcamo et al. 1995). PDE7 was originally cloned from a glioblastoma cDNA library (Michaeli et al. 1993) and is seen in the brain by Northern blot analysis (Bloom and Beavo, 1996). Even the very recently described PDE9 was shown by Northern blot analysis to be expressed with brain tissue (Soderling et al. 1998). Since the brain is such a complex tissue in many different subsets of neurons and glial cells it is important to determine where each of these isoforms are expressed in order to establish a role for these PDEs in neuronal function.

The expression of the cGMP-stimulated PDE in brain has been determined by activity measurements (Beavo et al. 1971; Murashima et al. 1990; Whalin et al. 1988), Northern analysis (Rosman et al. 1997; Sonnenburg et al. 1991), ribonuclease protection assay (Sonnenburg et al. 1991) and *in situ* hybridizations (Repaske et al. 1993). These data show that the PDE2 mRNA is found at high concentrations in the habenula, olfactory cortices, hippocampus, basal ganglia and cortex. Low levels were found in the hindbrain and spinal cord.

To date, no data on the distribution of PDE2 protein in the brain have been reported. Localization of protein in brain is particularly important since neurons extend dendrites and axons some distance from their cell bodies and distribution of the mRNA often does not reveal this subcellular distinction. A determination of where the protein is located within a neuron could help significantly in determining the physiological roles of these proteins.

In addition, the expression of guanylyl cyclases, both the soluble and membrane forms, is widespread throughout the brain. However, targets for cGMP action, such as cGMP-dependent protein kinase (Schlichter et al. 1980) and cyclic nucleotide gated channels (Bradley et al. 1997) have a much more limited localization pattern. Therefore PDE2, as a target for cGMP, may be an important component of the signaling pathways in many of these regions.

Immunohistochemistry using antiserum developed against the whole PDE2 protein shows that PDE2 expression is found in specific cells in almost all regions of the brain. The majority of PDE2 labeled cells appear to be neurons based on their morphological characteristics, including large soma, dendritic processes and axonal projections, and on their neuroanatomical location. The PDE2 protein is expressed in cell bodies as well as the axons and dendrites of many neurons.

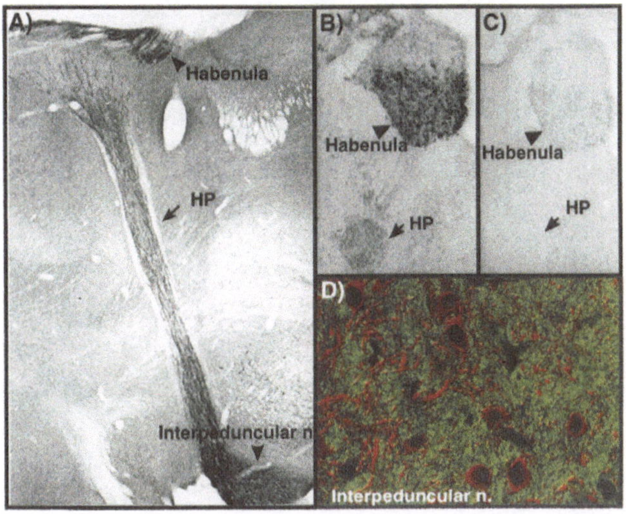

Fig. 6A–D. Expression of PDE2 protein in the habenula of the mouse brain. Immunohistochemistry was performed on perfused mouse brain floating sections. **A** Sagittal section showing PDE2 labeled cell bodies in the habenula with axons projecting to the interpeduncular n. (IPN). **B** Coronal section through the habenular n. showing PDE2 labeled neurons in the lower half of the habenula. **C** Control with protein competition of PDE2 antisera. **D** Double label immunofluorescence with PDE2 (*green*) and MAP-2 (*red*) antisera viewed with the confocal microscope. There is no overlap of signal suggesting that PDE2 is not expressed in neuronal cell bodies of the IPN and is probably localized to axon terminals originating from the habenula-peduncular tract (HP)

Two regions with the highest levels of PDE2 expression include the habenula (Fig. 6) and the basal ganglia. This is consistent with the mRNA localization studies reported previously (Repaske et al. 1993). Immunohistochemical studies have also shown that there is intense labeling of the axons as well as the cell bodies that can be seen projecting to their sites of termination in other regions of the brain. An example of axonal expression of PDE2 in habenular neurons is shown in Fig. 5. The habenula is a small structure that consists of small densely packed cells (Fig. 5B) located in the middle of the brain with long axons (Fig. 5A) that project to the interpeduncular nucleus (IPN) on the ventral surface. There is also immunoreactive PDE2 in the IPN. To determine if this signal is due to axon terminals from the habenular neurons or to cell bodies located in this nucleus, we double labeled the tissue with antisera against PDE2 and a monoclonal antibody for microtubule associated protein (MAP-2). MAP-2 is a protein expressed only

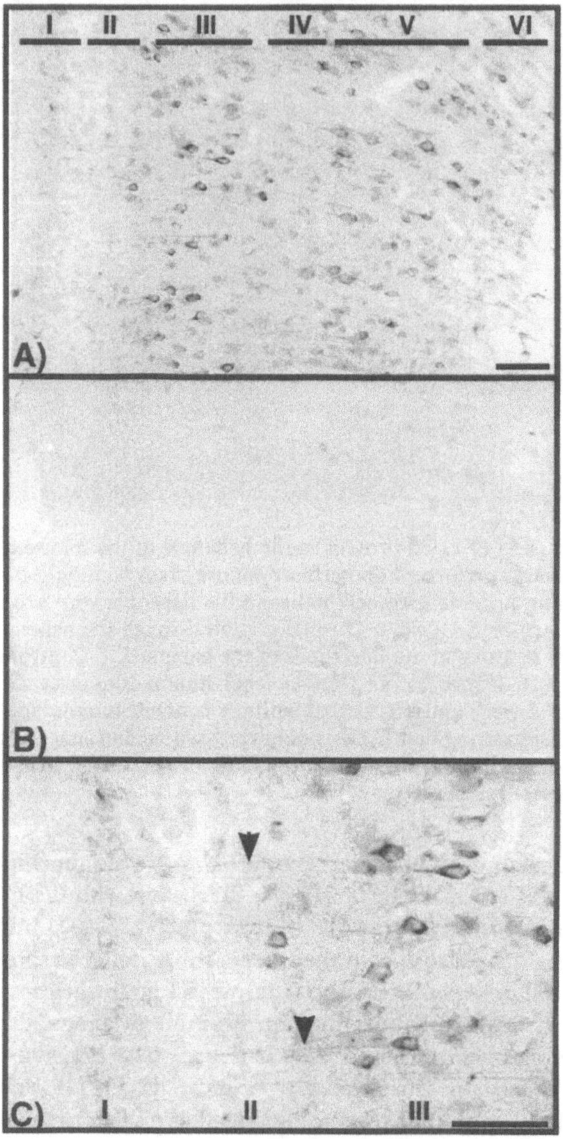

Fig. 7A–C. Expression of PDE2 protein in the cortex of the mouse brain. Immunohistochemistry was performed on perfused mouse brain floating sections. **A** Image showing all layers of the cortex within the frontal region of the brain. The Roman numerals above indicate specific neuronal layers. **B** Protein competition control in the same region as A). **C** Higher magnification showing only neuronal layers I-III. Large PDE2 labeled cell bodies are seen in layer III with their dendrites extending towards layer I and the surface of the brain. Scalebar= 50 μm

in cell bodies and dendrites. The data shows that all the immunoreactivity in this region is associated with axons and perhaps axonal terminals, rather than cell bodies in this region, since no co-localization with MAP-2 is apparent. This is consistent with lack of mRNA labeling by *in situ* hybridization studies (data not shown).

Dendrites are also commonly labeled by PDE2 antisera. This is most clearly seen in neurons of the cerebral cortex and the hippocampus. An example of dendritic PDE2 expression is shown in Fig. 7. PDE2 is expressed in neurons throughout the cerebral cortex. Immunohistochemistry (Juilfs, unpublished) and *in situ* hybridizations (Repaske et al. 1993) show that cell bodies in layers 3 and 5 express the highest concentration of PDE2. The immunohistochemical labeling shows that these cells consist of large triangular shaped cell bodies with dendrites that extend from the cell bodies towards layer I near the surface of the brain. This is consistent with these neurons being pyramidal in nature. PDE2 appears to be expressed throughout the cell bodies and dendrites; however, it is unclear whether the axons from these pyramidal cells also express PDE2.

Several areas, in addition to those described previously, have been found to express PDE2 protein. It had been reported (Repaske et al. 1993) that there is little or no PDE2 mRNA expression in the thalamic or hypothalamic nuclei, cerebellum or the midbrain. Immunohistochemistry, however, reveals that there is PDE2 expression in each of these areas. For example, Golgi cells in the granule cell layer of the cerebellum express PDE2 (Fig. 8). In relation to the number of cells in the cerebellum these represent a minor cell population, however, signal intensity in each neuron is high suggesting that there is a significant level of PDE2 protein expressed in these cells. The Golgi cells are interneurons involved in modulating input to the Purkinje cells from the mossy fiber pathway. In addition, the anterior group of the thalamic nuclei and select neurons of the suprachiasmatic nuclei of the hypothalamus express PDE2 protein.

There are also regions of the brain that express PDE2 that would not be detected by *in situ* hybridization studies. For example, the interpeduncular nucleus (as described previously) and the substantia nigra are considered to be part of the midbrain where low levels of PDE2 mRNA were detected. However, both of these structures express high levels of PDE2 protein. This signal appears to be present primarily in axons or axon terminals from cell bodies located in other areas of the brain, the habenula and the caudate putamen, respectively, and therefore the PDE2 message is not represented at high levels in these regions.

PDE2 is present in many of the brain structures that belong to the limbic system of the brain including the olfactory cortex, amygdala, hippocampal formation and the anterior thalamic nucleus. These structures are highly interconnected and are thought to be involved in the complex neural processes of learning, memory, and emotion consistent with a putative role in synaptic plasticity.

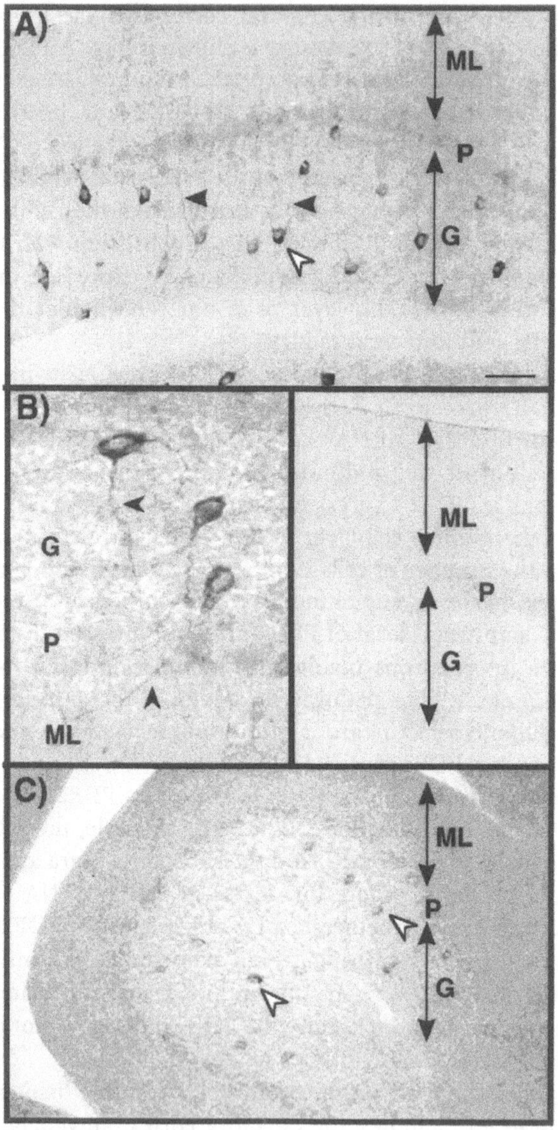

It is not clear at this time which isoforms of PDE2 are present in specific areas of the brain since neither the PDE2 antisera nor the riboprobes used in the localization studies were isoform specific. As described earlier three isoforms of PDE2 have been identified; both PDE2A2 and PDE2A3 are known to be expressed in the brain, and are thought to be differentially expressed at the subcellular level. Therefore, it is possible that more than one isoform is present in a single cell type. For example, one isoform might be specifically localized to the axons of neurons while the other might be restricted to the cell soma or dendrites. cGMP-stimulated PDE activity has been shown to be localized to the vesicles of clathrin coated pits in bovine brain (Silva and Puszkin, 1990) suggesting PDE2 could be involved in cAMP regulation of endocytosis or protein transport.

Finally, activation of PDE2 by cGMP has been shown to regulate cAMP metabolism in two cultured neuronal cell lines, PC12 cells (Whalin et al. 1990) and the neuroblastoma cell line NB-OK-1 (Delporte et al. 1996). These cell lines, therefore, might prove to be useful as tools for determining the role of PDE2 in neuronal cell function.

3
Cyclic GMP-Inhibited PDEs (PDE3)

As the name implies the cGMP-inhibited PDEs (PDE3s) have the distinguishing characteristic of being sensitive to inhibition of cAMP hydrolysis by cGMP. The PDE3 isoforms have been reviewed recently (Degerman et al. 1997; Manganiello et al. 1995). PDE3 has been purified from many tissues including heart (Harrison et al. 1986b), platelets (Grant and Colman, 1984) and adipocytes (Degerman et al. 1987). Although the cGMP-inhibited PDE (PDE3) is regulated by cGMP it does not have a distinct allosteric cGMP binding domain comparable to that of PDE2, PDE5, or PDE6. The catalytic domain of PDE3 has a higher affinity for cGMP than cAMP with K_ms of 0.1

Fig. 8A–D. Expression of PDE2 protein and mRNA in Golgi cells of the mouse cerebellum. **A** Immunohistochemistry was performed on perfused mouse brain floating sections. PDE2 immunoreactivity is seen in Golgi cells of the granule cell layer. White arrowhead points to a labeled cell body. Black arrowheads point to labeled dendrites extending towards the Purkinje cell layer. **B** Higher magnification of Golgi cells within the granule cell layer. Arrowheads point to dendritic processes. **C** Protein competition control. **D** *In situ* hybridization showing localization of PDE2 mRNA in cells with a similar distribution pattern as seen by immunohistochemistry. White arrowheads point to labeled cells. ML = molecular layer, P = Purkinje cell layer, G = granule cell layer

and 0.8 µM. However, it has as much as a 10-fold lower V_{max} for cGMP, and therefore interaction by cGMP is thought to inhibit cAMP hydrolysis by competition at the catalytic site (Harrison et al. 1986a). PDE3s are also characterized by a large number of selective inhibitors including milrinone, cilostazol and enoximone. Milrinone has been used as a cardiotonic agent (Varriale and Ramaprasad, 1997).

Two isoforms of PDE3 originating from distinct genes have been identified. PDE3A was cloned from a human cardiac cDNA library (Meacci et al. 1992) while PDE3B was cloned from a rat adipocyte cDNA library (Taira et al. 1993). Both isoforms have a 44 amino acid insert in the catalytic domain that is unique from other PDE families. A function for this domain has not yet been established although its presence is required for catalytic activity (Tang et al. 1997). PDE3A and PDE3B are greatly divergent at their N-termini with a proline rich region in the PDE3B protein. These regions do not appear to have any effect on cAMP hydrolysis or inhibitor sensitivity, however, PDE3A may be more sensitive to inhibition by cGMP (Leroy et al. 1996).

PDE3A and PDE3B are found in both the cytosolic and particulate fractions of tissue homogenates. However, when PDE3A and PDE3B are transfected into murine fibroblasts they are expressed exclusively in the particulate fraction (Leroy et al. 1996). This association is thought to occur by a putative transmembrane, hydrophobic domain found in the N-terminal region of the protein. When this region is truncated both proteins become cytosolic. This suggests either that in endogenous tissue proteolysis may occur to release the membrane bound forms into the cytosol or that there may be other splice variants of the PDE3 that code for these cytosolic isoforms.

The PDE3 proteins are difficult to isolate and susceptible to proteolysis. As a result the reported molecular weights for these proteins have ranged from 60–135 kDa. The expressed full length proteins, on the other hand, run at their predicted sizes of 123-125 kDa on SDS PAGE (Leroy et al. 1996). Several other PDE3s have been cloned. A truncated form of the PDE3 human platelet isoform was shown to originate from the same gene as the cardiac form (PDE3A) based on nucleotide sequence (Tang et al. 1997). The endogenous proteins differ, however, in that the platelet isoform is 110 kDa and cytosolic while the cardiac form is 126 kDa and primarily membrane associated. A placental form (Kasuya et al. 1995) that is also identical to the C-terminus of the cardiac form has been cloned and two transcript sizes (4.4 kb & 7.6 kb) in the placenta were identified by Northern blot analysis. The shorter form (74 kDa) was assigned a start site at nucleotide 1292 of the cardiac cDNA which corresponds to the putative exon 3. The two cDNAs exhibited differences when expressed in Sf9 cells. The shorter form is soluble while the longer form was associated with the membrane. They also have different affinities for cGMP. The 74 kDa form has a Kd from 3-20 μM and the longer form has a Kd 0.2–0.3 μM. PDE3 clones have also been identified in HEL cell lines (Cheung et al. 1996). These cell lines have similar protein expression pattern as platelets. These cells also have two transcript sizes similar to those seen in the placenta and a clone that corresponds to the cardiac form. This data might suggest that alternative start sites in the mRNA dictate whether the PDE3 is soluble or membrane associated.

PDE3A and PDE3B are differentially expressed in a tissue and cell type specific manner. Northern blot analysis shows that PDE3A is most abundant in heart, smooth muscle and platelets while PDE3B is found in both white and brown adipose cells, hepatocytes, renal collecting duct and developing spermatocytes (Reinhardt et al. 1995). Recent *in situ* hybridizations also show that both isoforms are expressed in the developing rat brain, however, their expression patterns are quite distinct (Reinhardt and Bondy, 1996). PDE3B mRNA is constitutively expressed throughout development in most regions of the brain while PDE3A mRNA is heterogenously expressed in different neurons at differing stages of development with little expression in the adult brain.

PDE3 activity has also been shown to be regulated by phosphorylation. PDE3A (Grant et al. 1988; Macphee et al. 1988) and PDE3B (Rascon et al. 1994) activity is increased when phosphorylated by cAMP-dependent protein kinase. This may function in a feedback loop where increased PDE activity by phosphorylation results in increased turnover of cAMP and decreased PKA activity.

PDE3B has been shown to be activated by insulin in adipocytes and hepatocytes by a phosphorylation event (Degerman et al. 1990; Smith et al. 1991). PDE3B is also expressed in pancreatic beta cells. In the pancreatic beta cells IGF-1 has been shown to stimulate PDE3B activity by phosphorylation leading to downregulation of insulin secretion (Zhao et al. 1997). This effect, however, does not appear to be mediated by cGMP.

Like PDE2, PDE3 may be involved in crosstalk between cAMP and cGMP pathways. However, unlike PDE2, PDE3 would play a role in those functions where cGMP and cAMP act synergistically. For example, a recent study showed that decreasing cGMP by inhibiting NO synthesis attenuates the renin secretory response to beta adrenergic stimulation in rabbits (Chiu et al. 1996). The decrease in cGMP allows PDE3 to become more active thereby decreasing cAMP and the secretory response. Milrinone increased renin activity and prevented suppression by cGMP verifying the participation of PDE3 in this physiological effect.

As stated in an earlier section PDE3 is the major cAMP hydrolyzing enzyme in platelets. Adenylyl cyclase activators and nitric oxide donors act synergistically to increase cAMP in human platelets (Fisch et al. 1995). The basis for this synergy has been shown to be modulation of cAMP levels by cGMP inhibition of PDE3 (Maurice and Haslam, 1990). In addition, the PDE3 inhibitor cilostazol decreases human platelet aggregation induced by shear stress (Minami et al. 1997) suggesting inhibition of PDE3 may be important therapeutically for the treatment or prevention of arterial occlusive diseases.

4
Cyclic GMP-Binding cGMP-Specific PDEs (PDE5)

PDE5 has been studied most extensively in the lung where it is the major cGMP binding protein. It has been partially purified from rat (Francis and Corbin, 1988), bovine (Thomas et al. 1990a), and guinea pig (Burns et al. 1992) lung. The PDE5 hydrolyzes only cGMP efficiently. It is composed of two identical 93 kDa subunits. Each subunit contains a non-catalytic cGMP binding domain similar to those found in PDE2 and PDE6. When cGMP is bound to this region PDE5 can be more readily phosphorylated by cGMP dependent protein kinase and cAMP dependent protein kinase. It is unclear how phosphorylation regulates PDE5 since no change in activity of the purified enzyme is apparent (Thomas et al. 1990b). In partially purified preparations, however, phosphorylation by PKA increases the V_{max} of cGMP hydrolysis (Burns et al. 1992). This data suggests that there may be other proteins involved in partially purified preparations that regulate the phosphorylation mediated increase in PDE5 hydrolytic activity.

Recent data shows that the inhibitory gamma subunit of the PDE6 complex can interact with PDE5 and prevent activation by phosphorylation of the partially purified PDE5 from guinea pig lung (Lochhead et al. 1997). In addition, two small proteins with molecular weights of 14 and 18 kDa are recognized by antisera against the PDE6 gamma subunit in lung and airway smooth muscle where the gamma subunit has not been observed, suggesting these are antigenically related proteins. These small proteins are phosphorylated by an endogenous kinase that appears to be regulated by a pertussis toxin sensitive G-protein. It appears the activity of PDE5 may be regulated in a manner more similar to that of PDE6 than that of PDE2.

To date two isoforms of PDE5 have been identified. A PDE5 cDNA was cloned from a bovine lung library (McAllister et al. 1993). Recently, a rat isoform was also cloned and appears to have a unique N-terminus (Kotera et

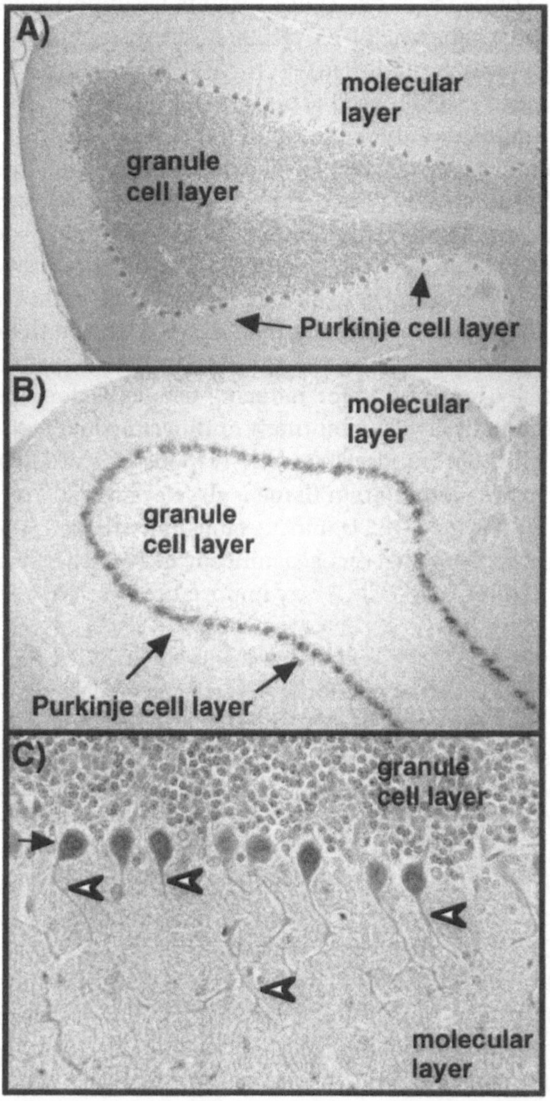

Fig. 9A–C. Expression of PDE5 protein and mRNA in Purkinje cells of the mouse cerebellum. **A** Immunohistochemistry using perfused paraffin embedded mouse brains. Blue = PDE5 immunoreactivity, Red = nuclear fast red. PDE5 is localized specifically to Purkinje cells (*arrows*). **B** *In situ* hybridization using digoxigenin labeled riboprobes specific for PDE5. The localization of PDE5 mRNA is also restricted to the Purkinje cells (*arrows*). C Higher magnification showing that PDE5 is expressed in the Purkinje cell dendritic tree (*arrowheads*) as well as in the cell bodies

al. 1997). The presence of two splice variants is supported by work done in this laboratory where two PDE5 variants have been cloned from a mouse lung library (Burns, unpublished).

PDE5 has also recently been identified in brain tissue. Its expression appears to be limited to only the Purkinje cell layer of the rat brain by *in situ* hybridization (Kotera et al. 1997). No expression was seen by *in situ* hybridizations or Northern analysis in other parts of the brain. Immunohistochemistry as well as *in situ* hybridizations from this laboratory demonstrate the expression of PDE5 in the same cell type (Burns et al. 1996) (Fig. 9). The PDE5 protein appears to be expressed in the cell bodies and the major branches of the extensive dendritic tree (Fig. 9C) known to be characteristic of the Purkinje neurons. Increases in cGMP in the cerebellum appear to be at least partially responsible for inducing cerebellar long-term depression (LTD). Inhibition of PDE5 by zaprinast or dipyrimadole both induce LTD in the parallel fiber-Purkinje cell synaptic responses (Hartell, 1996). Interestingly, PDE5 expression in brain tissue is developmentally regulated (Kotera et al. 1997). mRNA for PDE5 is undetectable in newborn rat brains, however Northern Blot analysis suggests a significant increase between day four and day 28. The regulation of cGMP may important in early rat cerebellar development.

Recent evidence suggests that the levels of PDE5 as well as other PDE isoforms are increased in pulmonary arteries in hypertensive rats (Hanson et al. 1998; MacLean et al. 1997). This is consistent with the use of PDE5 inhibitors as vasodilators (Cohen et al. 1996). Treatment of pulmonary hypertension with the PDE5 inhibitor zaprinast decreases pulmonary arterial pressure in fetal lambs (Skimming et al. 1996). Therefore, inhibition of PDE5 has therapeutic potential for the treatment of pulmonary hypertenstion.

Sildenafil is a potent and competitive inhibitor of PDE5 activity. This drug has proved effective as a treatment for male erectile disorder (Boolell et al. 1996). PDE5 is the major PDE isoform expressed in the corpus cavernosum. Inhibition of PDE5 is thought to enhance penile erection by increasing cGMP levels, similar to nitric oxide, resulting in the relaxation of the vascular smooth muscle. To date no major deleterious side effects appear to occur as a result of sildenafil treatment despite the fact that PDE5 is expressed in a variety of different tissues.

5 Calcium/Calmodulin-Dependent PDEs (PDE1)

Three gene families for the calcium/calmodulin (Ca/CaM) stimulated PDEs have been described to date. Each of the gene families have different hydrolytic specificities for cAMP and cGMP. PDE1A and PDE1B prefer cGMP as a substrate and both have a lower K_m for cGMP than for cAMP. The PDE1C family is unique as it hydrolyzes cAMP and cGMP equally well and has essentially the same K_m for both cAMP and cGMP. PDE1A is phosphorylated *in vitro* by PKA and PDE1B is phosphorylated *in vitro* by CaM-Kinase II . In both cases phosphorylation decreases the sensitivity of PDE1A and PDE1B to activation by Ca/CaM (Florio et al. 1994; Hashimoto et al. 1989; Sharma, 1991). Phosphorylation of PDE1C has not been reported to date.

Two N-terminal splice variants of PDE1A have been cloned from bovine (Sonnenburg et al. 1993) and human (Loughney et al. 1996) tissues and these correspond to the 59 and 61 kDa CaM-PDEs purified from bovine heart, lung and brain (Hansen and Beavo, 1982; Sharma and Wang, 1986; Sharma et al. 1980). Only one isoform of PDE1B has been cloned from bovine (Bentley et al. 1992), rat, mouse (Repaske et al. 1992) and most recently from human brain (Yu et al. 1997) and this corresponds to the 63 kDa CaM-PDE. Multiple splice variants of the PDE1C (72 kDa CaM-PDE) gene have been cloned (Loughney et al. 1996; Yan et al. 1996). These splice variants differ in both their N- and C-termini.

The different PDE1 isoforms also have different sensitivities for stimulation by calcium, differential inhibitor profiles, and are differentially expressed (Yan et al. 1996). *In situ* hybridizations in rat brain show distinct distributions for each isoform (Yan et al. 1994). For example, in the cerebellum PDE1B1 is expressed in Purkinje cells. PDE1C1 is present in Purkinje and granule cells while PDE1C5 is expressed only in granule cells. These *in situ* hybridization studies, however, do not indicate where the protein is localized. It is possible that PDE1B1 and PDE1C1 are localized to distinct subcellular compartments of the Purkinje cells and are likely involved in different cellular functions. As stated earlier the cGMP-specific PDE5 is also localized to the Purkinje cells (Kotera et al. 1997). These localization patterns suggest that in brain spatial differences in PDE expression are utilized in order to take advantage of the functional and regulatory differences between the PDE proteins.

In cultured cerebellar granule cells and astroglia, cGMP appears to be predominantly hydrolyzed by a CaM-PDE (Agull'o and Garc'ia, 1997). It is known that elevation of calcium in these cultures in response to NMDA results in cGMP increases by activation of the calcium dependent NOS-I. It has also been shown that the CaM-PDE activity is increased by the elevated calcium levels and reduces cGMP levels (Baltrons et al. 1997). Therefore, the CaM-PDE activation may be important for cGMP turnover after NOS-I stimulation.

PDE1 isoforms may also participate in the regulation of smooth muscle proliferation. Recently, is has been shown that PDE1C is expressed only in proliferating human vascular smooth muscle cells in culture but not in quiescent aortic smooth muscle cells (Rybalkin et al. 1997). PDE1B, on the other hand is expressed only in non-proliferating human vascular smooth muscle. So there appears to be a switch from a cGMP hydrolyzing CaM-PDE to a cGMP and cAMP hydrolyzing CaM-PDE in proliferating cells. Interestingly, induction of PDE1C in proliferating SMCs is not seen in monkey or rat where PDE1B and PDE1A are the major CaM-PDEs, respectively. Based on these results the PDE1 isoforms may be important targets for treatment of atherosclerosis and restenosis. This data again demonstrates the importance of determining species specific expression of the PDE proteins.

6
PDE9: Novel Low Km cGMP Specific PDEs

In the past, initial efforts to identify new PDEs have been based on the biochemical isolation and purification of distinct PDE activities from various tissue sources or standard methods of molecular biology. Recently a new family of PDEs, PDE9, has been identified and characterized by a novel method. This approach utilized bioinformatics to isolate cDNAs from the database of expressed sequence tags (EST) (Boguski et al. 1993; Hillier et al. 1996) based solely on sequence homology to PDE sequences. In this manner two isoforms of family 9 were independently isolated (Fisher et al. 1998; Soderling et al. 1998). PDE9A1 was cloned from mouse and PDE9A2 was cloned from human cDNA sources without prior knowledge of their PDE substrate specificity or kinetics.

PDE9A1 and PDE9A2 are 93 percent identical at the amino acid level (see Fig. 10). Both PDE9A1 and 9A2 contain a homologous predicted start methionine (arrow, Fig. 10) and a 237 amino acid conserved domain towards the C-terminus (boxed residues, Fig. 10). The most striking difference between these two PDEs is that 9A2 contains an N-terminal 60 amino acid insert compared to 9A1. This difference is most likely explained by alternative splicing, although it is plausible that species differences between mouse and humans also could account for this divergence. The functional consequence of this 60 amino acid insert, if any, is unknown at this time. It is possible the sequence of PDE9 N-terminal to the catalytic domain may serve some unknown regulatory function, analogous to the regulatory properties conferred by the N-terminal regions of other PDEs (Beavo, 1995). The inclusion or exclusion of these 60 amino acids within this region may therefore

```
                                  Start methionine
                                          |
9A1--->   1 ...................AGAMGAGSSSYRPKAIYLDIDGRIQKVVFSK 28
                                  ||.||||||||||||||||||||:|||
9A2--->   1 APRGGWRRESTVKSPSAAAGRRMGSGSSSYRPKAIYLDIDGRIQKVIFSK 50

         29 YCNSSDIMDLFCIATGLPRNTTISLLTTDDAMVSIDPTMPANSERTPYKV 78
            ||||||||||||||||||||||||||||||||||||||||||||||||||
         51 YCNSSDIMDLFCIATGLPRNTTISLLTTDDAMVSIDPTMPANSERTPYKV 100

         79 RPVAVKQVS......................................... 87
            ||||:||.|
        101 RPVAIKQLSAGVEDKRTTSRGQSAERPLRDRRVVGLEQPRREGAFESGQV 150

         88 ...................EREELIQGVLAQVAEQFSRAFKINELKAEVA 118
                               ||||||| |||||||||||||||||||||||
        151 EPRPREPQGCYQEGQRIPPEREELIQSVLAQVAEQFSRAFKINELKAEVA 200

        119 NHLAVLEKRVELEGLKVVEIEKCKSDIKKMREELAARNSRTNCPCKYSFL 168
            |||||||||||||||||||||||||||||||||||||.||||||||||||
        201 NHLAVLEKRVELEGLKVVEIEKCKSDIKKMREELAARSSRTNCPCKYSFL 250

        169 DN.KKLTPRRDVPTYPKYLLSPETIEALRKPTFDVWLWEPNEMLSCLEHM 217
            || |||||||||||||||||||||||||||||||||||||||||||||||
        251 DNHKKLTPRRDVPTYPKYLLSPETIEALRKPTFDVWLWEPNEMLSCLEHM 300

        218 YHDLGLVRDFSINPITLRRWLLCVHDNYRNNPFHNFRHCFCVTQMMYSMV 267
            ||||||||||||||:|||||| |||||||||||||||||||| |||||||
        301 YHDLGLVRDFSINPVTLRRWLFCVHDNYRNNPFHNFRHCFCVAQMMYSMV 350

        268 WLCGLQEKFSQMDILVLMTAAICHDLDHPGYNNTYQINARTELAVRYNDI 317
            ||| ||||||| |||:||||||||||||||||||||||||||||||||||
        351 WLCSLQEKFSQTDILILMTAAICHDLDHPGYNNTYQINARTELAVRYNDI 400

        318 SPLENHHCAIAFQILARPECNIFASVPPEGFRQIRQGMITLILATDMARH 367
            |||||||||:|||||| ||||||..:||:||:||||||||||||||||||
        401 SPLENHHCAVAFQILAEPECNIFSNIPPDGFKQIRQGMITLILATDMARH 450

        368 AEIMDSFKEKMENFDYSNEEHLTLLKMILIKCCDISNEVRPMEVAEPWVD 417
            |||||||||||||||||||||:||||||||||||||||||||||||||||
        451 AEIMDSFKEKMENFDYSNEEHMTLLKMILIKCCDISNEVRPMEVAEPWVD 500

        418 CLLEEYFMQSDREKSEGLPVAPFMDRDKVTKATAQIGFIKFVLIPMFETV 467
            ||||||||||||||||||||||||||||||||||||||||||||||||||
        501 CLLEEYFMQSDREKSEGLPVAPFMDRDKVTKATAQIGFIKFVLIPMFETV 550

        468 TKLFPVVEETMLRPLWESREHYEELKQLDDAMKELQKKTESLTSGAPENT 517
            |||||.||| ||.||||||: |||||.:|||||||||||:|||||| | .
        551 TKLFPMVEEIMLQPLWESRDRYEELKRIDDAMKELQKKTDSLTSGATEKS 600

        518 TEKNRDAKDSEGHSPPN  535
             |:.|| |.|||
        601 RERSRDVKNSEGDCA... 615
```

Fig. 10. Comparison of the amino acid sequence of mouse 9A1 and human 9A2. Predicted start methionine is marked by arrow. The C-terminal conserved domain found in all mammalian phosphodiesterases is boxed. These two isoforms are 93 percent identical at the amino acid level. PDE9A2 contains a 60 amino acid insert compared to PDE9A1. PKA and PKC consensus phosphorylation sites found within this amino acid insert are underlinded

modify the regulatory properties, if any, of this domain. Within this 60 amino acid stretch of sequence the only motifs found are a PKA phosphorylation site (KRTT) and a PKC phosphorylation consensus site (TSR) (underlined residues, Fig. 10). At this time, however, it is unknown if 9A2 is phosphorylated at these sites.

PDE9 has a very low Km for cGMP. Comparison of the predicted amino acid sequence of the conserved domain of PDE9 to the conserved domains of previously described mammalian PDEs shows PDE9 has between 29–36 percent sequence identity to other PDEs within this domain, including the other cGMP specific families PDE5 and PDE6. As PDEs within a family typically have greater than 85 percent sequence identity within the conserved domain, this demonstrates PDE9 represents a unique PDE family. PDE9 also has distinct kinetic properties from the previously known PDEs. Of the PDEs which can hydrolyze cGMP, PDE1 has a Km of 1–5 μM, PDE2 has a Km of 10 μM, PDE5 has a Km of 4 μM, and PDE6 has a Km of 17 μM; PDE9 has a Km of approximately 0.07 μM for cGMP (Fisher et al. 1998; Soderling et al. 1998). This is the lowest Km for cGMP for any PDE known to date. PDE9 is also very specific for the hydrolysis of cGMP, with a Km for cAMP about 3,000 times higher (230 μM (Fisher et al. 1998)) than that for cGMP. While PDE9 is similar to PDE5 and PDE6 in that cGMP hydrolysis is not inhibited effectively by high concentrations of cAMP, it is unique in that it does not contain a non-catalytic cGMP binding domain found in PDE5, PDE6, and PDE2.

Although PDE9 demonstrates cGMP specific PDE activity, this PDE is not inhibited well by most of the PDE inhibitors, including the non-specific inhibitor IBMX (IC50 greater than 200 μM). It is interesting to note that this PDE is also not inhibited by the PDE5 specific inhibitor sildenafil (Soderling et al. 1998), the PDE5 and PDE6 inhibitor dipyridamole, or the PDE2 inhibitor EHNA (Fisher et al. 1998; Soderling et al. 1998) at doses within ranges considered specific. Zaprinast, another PDE5 and PDE6 inhibitor will inhibit PDE9 at moderately high concentrations (29 μM for PDE9 versus 0.76–0.15 μM for PDE5/6) and PDE9 is inhibited by the relatively new PDE1 and PDE5 inhibitor SCH 51866 with an IC_{50} of 1.55 μM (Soderling et al. 1998). Inhibition of PDE9 by SCH 51866 is only 15 times higher than the IC_{50} of SCH 51866 for both PDE1 and 5 which both have an IC_{50} of approximately 0.1 μM (Vemulapalli et al. 1996). As SCH 51866 is the most potent inhibitor found so far for PDE9, SCH 51866 may also point to possible starting compounds for future studies assaying for inhibitors of PDE9 with greater selectivity and sensitivity. Also, SCH 51866 may be useful in future studies of PDE9 function in tissues or cells which have low or no endogenous PDE1 or PDE5 phosphodiesterase activity.

In summary, family 9 represents a very new and unique family of phosphodiesterases, dedicated to the specific low Km hydrolysis of cGMP. To date limited PDE9 expression distribution studies have been done in mouse and man , and these studies have not elucidated any single obvious physiologic function that PDE9 may regulate. A more detailed description of the tissue, cellular, and subcellular localization of PDE9 will likely follow the initial description and characterization of this family. It is anticipated that PDE9 will play a novel role in modulating intracellular cGMP levels to regulate physiologic responses.

Note: An eighth family of PDEs has been recently independently identified by both Soderling, S.H. and Fisher, D.A., et al.; Pfizer Central Research.

7
Conclusion

Because the PDEs are regulated by different mechanisms and because they are specifically expressed between and within cell types, they are considered to be important regulatory components in many signal transduction pathways. Therefore, identification and characterization of PDE isoform expression in specific cell types is important in elucidating these pathways. In addition, PDEs have become important targets for use in the therapeutic treatment of diverse diseases (Beavo, 1995; Hughes et al. 1997). There is a need, however, to design drugs specific for the increasing number of PDE isoforms in order to decrease the side effects that occur when multiple isoforms are targets for a single drug. In addition, it will be necessary to have an understanding of where each isoform is expressed, at the cellular and subcellular level, to aid in predicting the potential side effects that could occur upon drug treatment.

References

Agull'o, L., and Garc'ia, A.: Ca2+/calmodulin-dependent cyclic GMP phosphodiesterase activity in granule neurons and astrocytes from rat cerebellum. Eur J Pharmacol 323 (1):119–25, 1997

Bakalyar, H. A., and Reed, R. R.: Identification of a specialized adenylyl cyclase that may mediate odorant detection. Science 250 (4986):1403–6, 1990

Baltrons, M. A., Saadoun, S., Agullo, L., and Garcia, A.: Regulation by calcium of the nitric oxide cyclic GMP system in cerebellar granule cells and astroglia in culture. Journal Of Neuroscience Research 49 (3):333–341, 1997

Beavo, J. A.: Cyclic nucleotide phosphodiesterases:functional implications of multiple isoforms. Physiol Rev 75 (4):725–48, 1995

Beavo, J. A., Hardman, J. G., and Sutherland, E. W.: Stimulation of adenosine 3',5'-monophosphate hydrolysis by guanosine 3',5'-monophosphate. J Biol Chem. 246:3841–3846, 1971

Bentley, J. K., Kadlecek, A., Sherbert, C. H., Seger, D., Sonnenburg, W. K., Charbonneau, H., Novack, J. P., and Beavo, J. A.: Molecular cloning of cDNA encoding a "63"-kDa calmodulin-stimulated phosphodiesterase from bovine brain. J Biol Chem 267 (26):18676–82, 1992

Bloom, T. J., and Beavo, J. A.: Identification and tissue-specific expression of PDE7 phosphodiesterase splice variants. Proceedings Of The National Academy Of Sciences Of The United States Of America 93 (24):14188–14192, 1996

Boguski, M. S., Lowe, T. M., and Tolstoshev, C. M.: dbEST-database for "expressed sequence tags" [letter]. Nat Genet 4 (4):332–3, 1993

Boolell, M., Allen, M. J., Ballard, S. A., Gepi, A. S., Muirhead, G. J., Naylor, A. M., Osterloh, I. H., and Gingell, C.: Sildenafil: an orally active type 5 cyclic GMP-specific phosphodiesterase inhibitor for the treatment of penile erectile dysfunction. Int J Impot Res 8 (2):47–52, 1996

Bradley, J., Zhang, Y. N., Bakin, R., Lester, H. A., Ronnett, G. V., and Zinn, K.: Functional expression of the heteromeric "olfactory" cyclic nucleotide-gated channel in the hippocampus: A potential effector of synaptic plasticity in brain neurons. Journal Of Neuroscience 17 (6):1993–2005, 1997

Brunet, L. J., Gold, G. H., and Ngai, J.: General anosmia caused by a targeted disruption of the mouse olfactory cyclic nucleotide-gated cation channel. Neuron 17 (4):681–693, 1996

Buck, L., and Axel, R.: A novel multigene family may encode odorant receptors: a molecular basis for odor recognition. Cell 65 (1):175–87, 1991

Burns, F., Hanson, K., Miller, J. B., Rybalkin, S., Clarke, W. R., and Beavo, J. A.: Localization of PDE5 in Mouse Lung and Brain and Cloning of Two Mouse PDE5 Isoforms. Third International Conference on Cyclic Nucleotide Phosphodiesterases Abstract #67, 1996

Burns, F., Rodger, I. W., and Pyne, N. J.: The catalytic subunit of protein kinase A triggers activation of the type V cyclic GMP-specific phosphodiesterase from guinea-pig lung. Biochem J 283:487–491, 1992

Carcamo, B., Hurwitz, M. Y., Craft, C. M., and Hurwitz, R. L.: The mammalian pineal expresses the cone but not the rod cyclic gmp phosphodiesterase. Journal of Neurochemistry 65 (3):1085–1092, 1995

Charbonneau, H.: Structure-function relationships among cyclic nucleotide phosphodiesterases. In Cyclic nucleotide phosphodiesterases: structure, function, regulation and drug action, ed. by J. Beavo and M. D. Houslay, vol. 2, pp 267–298, John Wiley & Sons Ltd., Chichester, 1990

Charbonneau, H., Prusti, R. K., LeTrong, H., Sonnenburg, W. K., Mullaney, P. J., Walsh, K. A., and Beavo, J. A.: Identification of a noncatalytic cGMP-binding domain conserved in both the cGMP-stimulated and photoreceptor cyclic nucleotide phosphodiesterases. Proc Natl Acad Sci U S A 87 (1):288–92, 1990

Cherry, J. A., and Davis, R. L.: A mouse homolog of dunce, a gene important for learning and memory in drosophila, is preferentially expressed in olfactory receptor neurons. Journal of Neurobiology 28 (1):102–113, 1995

Cheung, P. P., Xu, H., McLaughlin, M. M., Ghazaleh, F. A., Livi, G. P., and Colman, R. W.: Human platelet cGI-PDE: expression in yeast and localization of the catalytic domain by deletion mutagenesis. Blood 88 (4):1321–9, 1996

Chiu, N., Park, I., and Reid, I. A.: Stimulation of renin secretion by the phosphodiesterase iv inhibitor rolipram. Journal of Pharmacology & Experimental Therapeutics 276 (3):1073–1077, 1996

Chiu, P. J. S., Vemulapalli, S., Chintala, M., Kurowski, S., Tetzloff, G. G., Brown, A. D., and Sybertz, E. J.: Inhibition of platelet adhesion and aggregation by E4021, a type V phosphodiesterase inhibitor, in guinea pigs. Naunyn Schmiedebergs Archives Of Pharmacology 355 (4):463–469, 1997

Cohen, A. H., Hanson, K., Morris, K., Fouty, B., Mcmurtry, I. F., Clarke, W., and Rodman, D. M.: Inhibition of cyclic 3'-5'-guanosine monophosphate-specific phosphodiesterase selectively vasodilates the pulmonary circulation in chronically hypoxic rats. Journal of Clinical Investigation 97 (1):172–179, 1996

Degerman, E., Belfrage, P., and Manganiello, V. C.: Structure, localization, and regulation of cGMP-inhibited phosphodiesterase (PDE3). Journal Of Biological Chemistry 272 (11):6823–6826, 1997

Degerman, E., Belfrage, P., Newman, A. H., Rice, K. C., and Manganiello, V. C.: Purification of the putative hormone-sensitive cyclic AMP phosphodiesterase from rat adipose tissue using a derivative of cilostamide as a novel affinity ligand. J Biol Chem 262 (12):5797–807, 1987

Degerman, E., Smith, C. J., Tornqvist, H., Vasta, V., Belfrage, P., and Manganiello, V. C.: Evidence that insulin and isoprenaline activate the cGMP-inhibited low-Km cAMP phosphodiesterase in rat fat cells by phosphorylation. Proc Natl Acad Sci U S A 87 (2):533–7, 1990

Delporte, C., Poloczek, P., and Winand, J.: Role of phosphodiesterase ii in cross talk between cgmp and camp in human neuroblastoma nb-ok-1 cells. American Journal of Physiology Cell Physiology 39 (1), 1996

Dickinson, N. T., Jang, E. K., and Haslam, R. J.: Activation of cGMP-stimulated phosphodiesterase by nitroprusside limits cAMP accumulation in human platelets: Effects on platelet aggregation. Biochemical Journal 323, 1997

Farber, D. B.: From mice to men: the cyclic GMP phosphodiesterase gene in vision and disease. The Proctor Lecture [published erratum appears in Invest Ophthalmol Vis Sci 1995 May;36(6):976]. Invest Ophthalmol Vis Sci 36 (2):263–75, 1995

Fisch, A., Michaelhepp, J., Meyer, J., and Darius, H.: Synergistic interaction of adenylate cyclase activators and nitric oxide donor sin-1 on platelet cyclic amp. European Journal of Pharmacology Molecular Pharmacology Section 289 (3):455–461, 1995

Fisher, D. A., Smith, J. F., Pillar, J. S., St.Denis, S. H., and Cheng, J. B.: Isolation and Characterization of PDE9A, a Novel Human cGMP-Specific Phosphodiesterase. J. Biol. Chem. In Press, 1998

Florio, V. A., Sonnenburg, W. K., Johnson, R., Kwak, K. S., Jensen, G. S., Walsh, K. A., and Beavo, J. A.: Phosphorylation of the 61-kDa calmodulin-stimulated cyclic nucleotide phosphodiesterase at serine 120 reduces its affinity for calmodulin. Biochemistry 33 (30):8948–8954, 1994

Francis, S. H., and Corbin, J. D.: Purification of cGMP-binding protein phosphodiesterase from rat lung. Methods Enzymol 159:722–9, 1988

Franks, D. J., and Macmanus, J. P.: Cyclic GMP stimulation and inhibition of cyclic AMP phosphodiesterase from thymic lymphocytes. Biochem. Biophys. Res. Commun. 42:844–849, 1971

Grant, P. G., and Colman, R. W.: Purification and characterization of a human platelet cyclic nucleotide phosphodiesterase. Biochemistry 23 (8):1801–1807, 1984

Grant, P. G., Mannarino, A. F., and Colman, R. W.: cAMP-mediated phosphorylation of the low-Km cAMP phosphodiesterase markedly stimulates its catalytic activity. Proc Natl Acad Sci U S A 85 (23):9071–5, 1988
Greer, C. A., Stewart, W. B., Teicher, M. H., and Shepherd, G. M.: Functional development of the olfactory bulb and a unique glomerular complex in the neonatal rat. J Neurosci 2 (12):1744–59, 1982
Han, X., Shimoni, Y., and Giles, W. R.: A cellular mechanism for nitric oxide-mediated cholinergic control of mammalian heart rate. Journal of General Physiology 106 (1):45–65, 1995
Han, X. Q., Kobzik, L., Zhao, Y. Y., Opel, D. J., Liu, W. D., Kelly, R. A., and Smith, T. W.: Nitric oxide regulation of atrioventricular node excitability. Canadian Journal Of Cardiology 13 (12):1191–1201, 1997
Hansen, R. S., and Beavo, J. A.: Purification of two calcium/calmodulin-dependent forms of cyclic nucleotide phosphodiesterase by using conformation-specific monoclonal antibody chromatography. Proc Natl Acad Sci U S A 79 (9):2788–92, 1982
Hanson, K. A., Ziegler, J. W., Rybalkin, S., Miller, J. W., Abman, S. H., and Clarke, J.: Chronic Pulmonary Hypertension Increases Fetal Lung cGMP Phosphodiesterase Activity. Am. J. Phy.: Lung cellular and molecular physiology In Press, 1998
Harrison, S. A., Reifsnyder, D. H., Gallis, B., Cadd, G. G., and Beavo, J. A.: Isolation and characterization of bovine cardiac muscle cGMP-inhibited phosphodiesterase: a receptor for new cardiotonic drugs. Mol Pharmacol 29:506–514, 1986a
Harrison, S. A., Reifsnyder, D. H., Gallis, B., Cadd, G. G., and Beavo, J. A.: Isolation and characterization of bovine cardiac muscle cGMP-inhibited phosphodiesterase: a receptor for new cardiotonic drugs. Mol Pharmacol 29:506–514, 1986b
Hartell, N. A.: Inhibition of cgmp breakdown promotes the induction of cerebellar long-term depression. Journal of Neuroscience 16 (9):2881–2890, 1996
Hartzell, H. C., and Fischmeister, R.: Opposite effects of cyclic GMP and cyclic AMP on Ca2+ current in single heart cells. Nature 323 (6085):273–5, 1986
Hashimoto, Y., Sharma, R. K., and Soderling, T. R.: Regulation of Ca2+/calmodulin-dependent cyclic nucleotide phosphodiesterase by the autophosphorylated form of Ca2+/calmodulin-dependent protein kinase II. J Biol Chem 264 (18): 10884–7, 1989
Haynes, J., Killilea, D. W., Peterson, P. D., and Thompson, W. J.: Erythro-9-(2-hydroxy-3-nonyl)adenine inhibits cyclic-3',5'-guanosine monophosphate-stimulated phosphodiesterase to reverse hypoxic pulmonary vasoconstriction in the perfused rat lung. Journal of Pharmacology & Experimental Therapeutics 276 (2):752–757, 1996
Hillier, L. D., Lennon, G., Becker, M., Bonaldo, M. F., Chiapelli, B., Chissoe, S., Dietrich, N., DuBuque, T., Favello, A., Gish, W., Hawkins, M., Hultman, M., Kucaba, T., Lacy, M., Le, M., Le, N., Mardis, E., Moore, B., Morris, M., Parsons, J., Prange, C., Rifkin, L., Rohlfing, T., Schellenberg, K., Marra, M., and et, a. l.: Generation and analysis of 280,000 human expressed sequence tags. Genome Res 6 (9):807–28, 1996
Hughes, B., Owens, R., Perry, M., Warrellow, G., and Allen, R.: PDE 4 inhibitors: The use of molecular cloning in the design and development of novel drugs. Drug Discovery Today 2 (3):89–101, 1997
Iona, S., Cuomo, M., Bushnik, T., Naro, F., Sette, C., Hess, M., Shelton, E. R., and Conti, M.: Characterization of the rolipram-sensitive, cyclic AMP-specific phos-

phodiesterases: Identification and differential expression of immunologically distinct forms in the rat brain. Molecular Pharmacology 53 (1):23–32, 1998

Ito, M., Nishikawa, M., Fujioka, M., Miyahara, M., Isaka, N., Shiku, H., and Nakano, T.: Characterization of the isoenzymes of cyclic nucleotide phosphodiesterase in human platelets and the effects of E4021. Cellular Signalling 8 (8):575–581, 1996

Juilfs, D. M., Fulle, H. J., Zhao, A. Z., Houslay, M. D., Garbers, D. L., and Beavo, J. A.: A subset of olfactory neurons that selectively express cGMP-stimulated phosphodiesterase (PDE2) and guanylyl cyclase-D define a unique olfactory signal transduction pathway. Proceedings Of The National Academy Of Sciences Of The United States Of America 94 (7):3388–3395, 1997

Kasuya, J., Goko, H., and Fujita, Y. Y.: Multiple transcripts for the human cardiac form of the cGMP-inhibited cAMP phosphodiesterase. J Biol Chem 270 (24):14305–12, 1995

Kishi, Y., Ashikaga, T., Watanabe, R., and Numano, F.: Atrial natriuretic peptide reduces cyclic AMP by activating cyclic GMP-stimulated phosphodiesterase in vascular endothelial cells. J Cardiovasc Pharmacol 24 (3):351–357, 1994

Kotera, J., Yanaka, N., Fujishige, K., Imai, Y., Akatsuka, H., Ishizuka, T., Kawashima, K., and Omori, K.: Expression of rat cGMP-binding cGMP-specific phosphodiesterase mRNA in Purkinje cell layers during postnatal neuronal development. Eur J Biochem 249 (2):434–42, 1997

Lee, M. A., West, R. J., and Moss, J.: Atrial natriuretic factor reduces cyclic adenosine monophosphate content of human fibroblasts by enhancing phosphodiesterase activity. J Clin Invest 82 (2):388–93, 1988

Leroy, M. J., Degerman, E., Taira, M., Murata, T., Wang, L. H., Movsesian, M. A., Meacci, E., and Manganiello, V. C.: Characterization of two recombinant pde3 (cgmp-inhibited cyclic nucleotide phosphodiesterase) isoforms, rcgip1 and hcgip2, expressed in nih 3006 murine fibroblasts and sf9 insect cells. Biochemistry 35 (31):10194–10202, 1996

Lobban, M., Shakur, Y., Beattie, J., and Houslay, M. D.: Identification of two splice variant forms of type-IVB cyclic AMP phosphodiesterase, DPD (rPDE-IVB1) and PDE-4 (rPDE-IVB2) in brain: selective localization in membrane and cytosolic compartments and differential expression in various brain regions. Biochem J, 1994

Lochhead, A., Nekrasova, E., Arshavsky, V. Y., and Pyne, N. J.: The regulation of the cGMP-binding cGMP phosphodiesterase by proteins that are immunologically related to gamma subunit of the photoreceptor cGMP phosphodiesterase. Journal Of Biological Chemistry 272 (29):18397–18403, 1997

Loughney, K., Martins, T. J., Harris, E. A. S., Sadhu, K., Hicks, J. B., Sonnenburg, W. K., Beavo, J. A., and Ferguson, K.: Isolation and characterization of cDNAs corresponding to two human calcium, calmodulin-regulated, 3',5'-cyclic nucleotide phosphodiesterases. Journal Of Biological Chemistry 271 (2):796–806, 1996

MacFarland, R. T., Zelus, B. D., and Beavo, J. A.: High concentrations of a cGMP-stimulated phosphodiesterase mediate ANP-induced decreases in cAMP and steroidogenesis in adrenal glomerulosa cells. J Biol Chem 266 (1):136–42, 1991

MacLean, M. R., Johnston, E. D., McCulloch, K. M., Pooley, L., Houslay, M. D., and Sweeney, G.: Phosphodiesterase isoforms in the pulmonary arterial circulation of the rat: Changes in pulmonary hypertension. Journal Of Pharmacology And Experimental Therapeutics 283 (2):619–624, 1997

Macphee, C. H., Reifsnyder, D. H., Moore, T. A., Lerea, K. M., and Beavo, J. A.: Phosphorylation results in activation of a cAMP phosphodiesterase in human platelets. J Biol Chem 263 (21):10353–8, 1988
Manganiello, V. C., Taira, M., Degerman, E., and Belfrage, P.: TYPE III CGMP-INHIBITED CYCLIC NUCLEOTIDE PHOSPHODIESTERASES (PDE 3 GENE FAMILY) [Review]. Cellular Signalling 7 (5):445–455, 1995
Manganiello, V. C., Tanaka, T., and Murashima, S.: Cyclic GMP-stimulated cyclic nucleotide phosphodiesterases. In Cyclic nucleotide phosphodiesterases: structure, regulation and drug action, ed. by J. Beavo and M. D. Houslay, vol. 2, pp 61–86, John Wiley & Sons, Chichester, 1990
Martins, T. J., Mumby, M. C., and Beavo, J. A.: Purification and characterization of a cyclic GMP-stimulated cyclic nucleotide phosphodiesterase from bovine tissues. J Biol Chem 257:1973–1979, 1982
Maurice, D. H., and Haslam, R. J.: Molecular basis of the synergistic inhibition of platelet function by nitrovasodilators and activators of adenylate cyclase: inhibition of cyclic AMP breakdown by cyclic GMP. Mol Pharmacol 37 (5):671–81, 1990
McAllister, L. L. M., Sonnenburg, W. K., Kadlecek, A., Seger, D., Trong, H. L., Colbran, J. L., Thomas, M. K., Walsh, K. A., Francis, S. H., Corbin, J. D., and Beavo, J. A.: The structure of a bovine lung cGMP-binding, cGMP-specific phosphodiesterase deduced from a cDNA clone. J Biol Chem 268 (30):22863–73, 1993
Mcphee, I., Pooley, L., Lobban, M., Bolger, G., and Houslay, M. D.: Identification, characterization and regional distribution in brain of rpde-6 (rnpde4a5), a novel splice variant of the pde4a cyclic amp phosphodiesterase family. Biochemical Journal , 1995
Meacci, E., Taira, M., Moos, M. J., Smith, C. J., Movsesian, M. A., Degerman, E., Belfrage, P., and Manganiello, V.: Molecular cloning and expression of human myocardial cGMP-inhibited cAMP phosphodiesterase. Proc Natl Acad Sci U S A 89 (9):3721–5, 1992
Mery, P. F., Lohmann, S. M., Walter, U., and Fischmeister, R.: Ca2+ current is regulated by cyclic GMP-dependent protein kinase in mammalian cardiac myocytes. Proc Natl Acad Sci U S A 88 (4):1197–201, 1991
Mery, P. F., Pavoine, C., Pecker, F., and Fischmeister, R.: Erythro-9-(2-hydroxy-3-nonyl)adenine inhibits cyclic gmp-stimulated phosphodiesterase in isolated cardiac myocytes. Molecular Pharmacology 48 (1):121–130, 1995
Michaeli, T., Bloom, T. J., Martins, T., Loughney, K., Ferguson, K., Riggs, M., Rodgers, L., Beavo, J. A., and Wigler, M.: Isolation and characterization of a previously undetected human cAMP phosphodiesterase by complementation of cAMP phosphodiesterase-deficient Saccharomyces cerevisiae. J Biol Chem 268 (17):12925–32, 1993
Michie, A. M., Lobban, M., Muller, T., Harnett, M. M., and Houslay, M. D.: Rapid regulation of PDE-2 and PDE-4 cyclic AMP phosphodiesterase activity following ligation of the T cell antigen receptor on thymocytes: analysis using the selective inhibitors erythro-9-(2-hydroxy-3-nonyl)-adenine (EHNA) and rolipram. Cell Signal 8 (2):97–110, 1996
Minami, N., Suzuki, Y., Yamamoto, M., Kihira, H., Imai, E., Wada, H., Kimura, Y., Ikeda, Y., Shiku, H., and Nishikawa, M.: Inhibition of shear stress-induced platelet aggregation by cilostazol, a specific inhibitor of cGMP-inhibited phosphodiesterase, in vitro and ex vivo. Life Sciences 61 (25), 1997

Miot, F., Van, H. P. J., and Erneux, C.: Specificity of cGMP binding to a purified cGMP-stimulated phosphodiesterase from bovine adrenal tissue. Eur J Biochem 149 (1):59–65, 1985

Moss, J., Manganiello, V. C., and Vaughan, M.: Substrate and effector specificity of a guanosine 3': 5'-monophosphate phosphodiesterase from rat liver. J Biol Chem 252 (15):5211–5, 1977

Murashima, S., Tanaka, T., Hockman, S., and Manganiello, V.: Characterization of particulate cyclic nucleotide phosphodiesterases from bovine brain: purification of a distinct cGMP-stimulated isoenzyme. Biochemistry 29 (22):5285–92, 1990

Podzuweit, T., Nennstiel, P., and Muller, A.: Isozyme selective inhibition of cgmp-stimulated cyclic nucleotide phosphodiesterases by erythro-9-(2-hydroxy-3-nonyl) adenine. Cellular Signalling 7 (7):733–738, 1995

Pyne, N. J., Cooper, M. E., and Houslay, M. D.: Identification and characterization of both the cytosolic and particulate forms of cyclic GMP-stimulated cyclic AMP phosphodiesterase from rat liver. Biochem J 234:325–334, 1986

Rascon, A., Degerman, E., Taira, M., Meacci, E., Smith, C. J., Manganiello, V., Belfrage, P., and Tornqvist, H.: Identification of the Phosphorylation Site in Vitro for cAMP-Dependent Protein Kinase on the Rat Adipocyte cGMP- Inhibited cAMP Phosphodiesterase. J Biol Chem 269 (16):11962–11966, 1994

Reinhardt, R. R., and Bondy, C. A.: Differential cellular pattern of gene expression for two distinct cgmp-inhibited cyclic nucleotide phosphodiesterases in developing and mature rat brain. Neuroscience 72 (2):567–578, 1996

Reinhardt, R. R., Chin, E., Zhou, J., Taira, M., Murata, T., Manganiello, V. C., and Bondy, C. A.: Distinctive anatomical patterns of gene expression for cgmp-inhibited cyclic nucleotide phosphodiesterases. Journal of Clinical Investigation 95 (4):1528–1538, 1995

Repaske, D. R., Corbin, J. G., Conti, M., and Goy, M. F.: A cyclic GMP-stimulated cyclic nucleotide phosphodiesterase gene is highly expressed in the limbic system of the rat brain. Neuroscience 56 (3):673–86, 1993

Repaske, D. R., Swinnen, J. V., Jin, S. L., Van, W. J. J., and Conti, M.: A polymerase chain reaction strategy to identify and clone cyclic nucleotide phosphodiesterase cDNAs. Molecular cloning of the cDNA encoding the 63-kDa calmodulin-dependent phosphodiesterase. J Biol Chem 267 (26):18683–8, 1992

Ressler, K. J., Sullivan, S. L., and Buck, L. B.: Information coding in the olfactory system: evidence for a stereotyped and highly organized epitope map in the olfactory bulb. Cell 79 (7):1245–55, 1994

Ring, G., Mezza, R. C., and Schwob, J. E.: Immunohistochemical identification of discrete subsets of rat olfactory neurons and the glomeruli that they innervate. J Comp Neurol 388 (3):415–34, 1997

RivetBastide, M., Vandecasteele, G., Hatem, S., Verde, I., Benardeau, A., Mercadier, J. J., and Fischmeister, R.: cGMP-stimulated cyclic nucleotide phosphodiesterase regulates the basal calcium current in human atrial myocytes. Journal Of Clinical Investigation 99 (11):2710–2718, 1997

Rosman, G. J., Martins, T. J., Sonnenburg, W. K., Beavo, J. A., Ferguson, K., and Loughney, K.: Isolation and characterization of human cDNAs encoding a cGMP-stimulated 3',5'-cyclic nucleotide phosphodiesterase. Gene 191 (1):89–95, 1997

Rybalkin, S. D., Bornfeldt, K. E., Sonnenburg, W. K., Rybalkina, I. G., Kwak, K. S., Hanson, K., Krebs, E. G., and Beavo, J. A.: Calmodulin-stimulated cyclic nucleotide phosphodiesterase (PDE1C) is induced in human arterial smooth muscle cells of the synthetic, proliferative phenotype. Journal Of Clinical Investigation 100 (10):2611–2621, 1997

Schlichter, D. J., Detre, J. A., Aswad, D. W., Chehrazi, B., and Greengard, P.: Localization of cyclic GMP-dependent protein kinase and substrate in mammalian cerebellum. Proc Natl Acad Sci U S A 77 (9):5537–41, 1980

Sharma, R. K.: Phosphorylation and characterization of bovine heart calmodulin-dependent phosphodiesterase. Biochemistry 30 (24):5963–8, 1991

Sharma, R. K., and Wang, J. H.: Purification and characterization of bovine lung calmodulin-dependent cyclic nucleotide phosphodiesterase. An enzyme containing calmodulin as a subunit. J Biol Chem 261:14160–14166, 1986

Sharma, R. K., Wang, T. H., Wirch, E. H., and J., W.: Purification and properties of bovine brain calmodulin-dependent cyclic nucleotide phosphodiesterase. J. Biol. Chem. 255:5916–5923, 1980

Shirayama, T., and Pappano, A. J.: Biphasic effects of intrapipette cyclic guanosine monophosphate on L-type calcium current and contraction of guinea pig ventricular myocytes. Journal Of Pharmacology And Experimental Therapeutics 279 (3):1274–1281, 1996

Silva, W. I., and Puszkin, S.: Equilibrium kinetics model for the cGMP-stimulated phosphodiesterase of brain coated vesicles. Bol Asoc Med P R 82 (9):407–11, 1990

Skimming, J. W., Demarco, V. G., Kadowitz, P. J., and Cassin, S.: Effects of zaprinast and dissolved nitric oxide on the pulmonary circulation of fetal sheep. Pediatric Research 39 (2):223–228, 1996

Smith, C. J., Vasta, V., Degerman, E., Belfrage, P., and Manganiello, V. C.: Hormone-sensitive cyclic GMP-inhibited cyclic AMP phosphodiesterase in rat adipocytes. Regulation of insulin- and cAMP-dependent activation by phosphorylation. J Biol Chem 266 (20):13385–90, 1991

Soderling, S. H., Bayuga, S. J., and Beavo, J. A.: Identification and Characterization of a Novel Family of Cyclic Nucleotide Phosphodiesterases. J. Bio. Chem. In Press, 1998

Sonnenburg, W. K., Mullaney, P. J., and Beavo, J. A.: Molecular cloning of a cyclic GMP-stimulated cyclic nucleotide phosphodiesterase cDNA. Identification and distribution of isozyme variants. J Biol Chem 266 (26):17655–61, 1991

Sonnenburg, W. K., Seger, D., and Beavo, J. A.: Molecular cloning of a cDNA encoding the "61-kDa" calmodulin-stimulated cyclic nucleotide phosphodiesterase. Tissue-specific expression of structurally related isoforms. J Biol Chem 268 (1):645–52, 1993

Stroop, S. D., and Beavo, J. A.: Structure and function studies of the cGMP-stimulated phosphodiesterase. J Biol Chem 266 (35):23802–9, 1991

Suttorp, N., Hippenstiel, S., Fuhrmann, M., Krull, M., and Podzuweit, T.: Role of nitric oxide and phosphodiesterase isoenzyme ii for reduction of endothelial hyperpermeability. American Journal of Physiology Cell Physiology 39 (3), 1996

Taira, M., Hockman, S. C., Calvo, J. C., Taira, M., Belfrage, P., and Manganiello, V. C.: Molecular cloning of the rat adipocyte hormone-sensitive cyclic GMP-inhibited cyclic nucleotide phosphodiesterase. J Biol Chem 268 (25):18573–9, 1993

Tang, K. M., Jang, E. K., and Haslam, R. J.: Expression and mutagenesis of the catalytic domain of cGMP-inhibited phosphodiesterase (PDE3) cloned from human platelets. Biochem J , 1997

Teicher, M. H., Stewart, W. B., Kauer, J. S., and Shepherd, G. M.: Suckling pheromone stimulation of a modified glomerular region in the developing rat olfactory bulb revealed by the 2-deoxyglucose method. Brain Res 194 (2):530–5, 1980

Thomas, M. K., Francis, S. H., and Corbin, J. D.: Characterization of a purified bovine lung cGMP-binding cGMP phosphodiesterase. J Biol Chem 265 (25):14964–70, 1990a

Thomas, M. K., Francis, S. H., and Corbin, J. D.: Substrate- and kinase-directed regulation of phosphorylation of a cGMP-binding phosphodiesterase by cGMP. J Biol Chem 265 (25):14971–8, 1990b

Trong, H. L., Beier, N., Sonnenburg, W. K., Stroop, S. D., Walsh, K. A., Beavo, J. A., and Charbonneau, H.: Amino acid sequence of the cyclic GMP stimulated cyclic nucleotide phosphodiesterase from bovine heart. Biochemistry 29 (44):10280–8, 1990

Varriale, P., and Ramaprasad, S.: Short-term intravenous milrinone for severe congestive heart failure: the good, bad, and not so good. Pharmacotherapy 17 (2):371–4, 1997

Vassar, R., Chao, S. K., Sitcheran, R., Nunez, J. M., Vosshall, L. B., and Axel, R.: Topographic organization of sensory projections to the olfactory bulb. Cell 79 (6):981–91, 1994

Vemulapalli, S., Watkins, R. W., Chintala, M., Davis, H., Ahn, H. S., Fawzi, A., Tulshian, D., Chiu, P., Chatterjee, M., Lin, C. C., and Sybertz, E. J.: Antiplatelet and antiproliferative effects of SCH 51866, a novel type 1 and type 5 phosphodiesterase inhibitor. Journal Of Cardiovascular Pharmacology 28 (6):862–869, 1996

Whalin, M. E., Strada, S. J., and Thompson, W. J.: Purification and partial characterization of membrane-associated type II (cGMP-activatable) cyclic nucleotide phosphodiesterase from rabbit brain. Biochim Biophys Acta 972 (1):79–94, 1988

Whalin, M. W., Strada, S. J., Scammell, J. G., and Thompson, J. G.: Regulation of cAMP metabolism in PC12 cells by type II (cGMP-activatable) cyclic nucleotide phosphodiesterase. In Purines in Cellular Signaling Targets for New Drugs, ed. by J. W. D. a. V. M. K.A. Jacobson, pp 323–328, Springer-Verlag, New York, 1990

Yamamoto, T., Manganiello, V. C., and Vaughan, M.: Purification and characterization of cyclic GMP-stimulated cyclic nucleotide phosphodiesterase from calf liver. J Biol Chem 258:12526–12533, 1983

Yan, C., Bentley, J. K., Sonnenburg, W. K., and Beavo, J. A.: Differential expression of the 61 kDa and 63 kDa calmodulin-dependent phosphodiesterases in the mouse brain. J. Neuroscience 14 (3):973–984, 1994

Yan, C., Zhao, A. Z., Bentley, J. K., and Beavo, J. A.: The calmodulin-dependent phosphodiesterase gene pde1c encodes several functionally different splice variants in a tissue-specific manner. Journal of Biological Chemistry 271 (41):25699–25706, 1996

Yan, C., Zhao, A. Z., Bentley, J. K., Loughney, K., Ferguson, K., and Beavo, J. A.: Molecular cloning and characterization of a calmodulin-dependent phosphodiesterase enriched in olfactory sensory neurons. Proc. Nat. Acad. Sci. U.S.A. 92 (21):9677–9681, 1995

Yang, Q., Paskind, M., Bolger, G., Thompson, W. J., Repaske, D. R., Cutler, L. S., and Epstein, P. M.: A novel cyclic GMP stimulated phosphodiesterase from rat brain. Biochem. Biophys. Res. Comm. 205:1850–1858, 1994

Yu, J., Wolda, S. L., Frazier, A. L. B., Florio, V. A., Martins, T. J., Snyder, P. B., Harris, E. A. S., McCaw, K. N., Farrell, C. A., Steiner, B., Bentley, J. K., Beavo, J. A., Ferguson, K., and Gelinas, R.: Identification and characterisation of a human calmodulin-stimulated phosphodiesterase PDE1B1. Cellular Signalling 9 (7):519–529, 1997

Zhao, A. Z., Zhao, H., Teague, J., Fujimoto, W., and Beavo, J. A.: Attenuation of insulin secretion by insulin-like growth factor 1 is mediated through activation of phosphodiesterase 3B. Proceedings Of The National Academy Of Sciences Of The United States Of America 94 (7):3223–3228, 1997

Structure and Function of cGMP-Dependent Protein Kinases

A. Pfeifer, P. Ruth, W. Dostmann, M. Sausbier, P. Klatt, and F. Hofmann

Institut für Pharmakologie und Toxikologie der TU, Biedersteiner Straße 29, D-80802 München, Germany

Contents

1 Introduction

Cyclic GMP is a widely distributed second messenger which was discovered more than 30 years ago. Cyclic GMP is synthesized by soluble and particulate guanylyl cyclases and is degraded by the specific phosphodiesterase V. The activity of the guanylyl cyclases is controlled by a broad spectrum of substances including nitric oxide (NO), organic nitrates, peptide hormones and toxins. In turn, cGMP regulates various physiological processes ranging from smooth muscle tone, neuronal excitability, epithelial electrolyte transport to phototransduction in the retina. The discovery of a cGMP-dependent protein kinase (cGMP kinase) activity first in lobster muscle (Kuo and Greengard 1970) and then in rat cerebellum (Hofmann and Sold 1972) led to the view that – in analogy to cAMP-dependent protein kinase (cAMP kinase) – cGMP kinase mediates the physiological functions of cGMP. Later it became clear that eukaryotic cells have at least three distinct classes of cGMP receptor proteins: cGMP regulated phosphodiesterases (PDEs) (Sonnenburg and Beavo 1994), cyclic nucleotide gated cation channels (CNGs) (Biel et al. see Chapter 5) and cGMP kinases. Initial biochemical studies and experiments in isolated cells and tissues indicated that cGMP kinase controls physiological functions by similar or identical pathways as cAMP kinases. It was also reported that cGMP kinase is activated by cAMP under physiological conditions suggesting a rather "nonspecific" function of cGMP kinases. More recent experiments which include inactivation of the cGMP kinase genes show that cGMP kinases very specifically regulate distinct cellular functions by pathways which are separate from those used by cAMP kinases. This review summarizes the biochemical, functional and physiological properties of the cGMP kinases.

2 cGMP Kinase Genes

cGMP kinase has been found in various eukaryotic organisms. A cGMP-dependent protein kinase activity was initially detected in arthropodes (Kuo and Greengard 1970) and has also been identified in other invertebrate species, such as *silkworm* (Takahashi et al. 1974), *Paramecium* (Miglietta and Nelson 1988), *Tetrahymena* (Murofushi 1974), and *Dictyostelium discoideum* (Wanner and Wurster 1990). After the initial description of a soluble, mammalian cGMP-stimulated kinase activity (Hofmann and Sold 1972), a membrane-bound cGMP kinase protein was isolated from intestinal epithelial cells (De Jonge 1981). This dual localization led to the hypothesis that two different forms of mammalian cGMP kinases exist, cGMP kinase I and cGMP kinase II.

The cloning of cGMP kinase I from bovine (Wernet et al. 1989) and human (Sandberg et al. 1989) smooth muscle revealed alternative splicing of the amino-terminus yielding cGMP kinase Iα and cGMP kinase Iß. Apart from the amino-terminus, the two enzymes are identical as indicated by the comparison of full-length clones of bovine cGMP kinase Iα (Wernet et al. 1989) and Iß (Ruth et al. 1997). More recently, cGMP kinase II was cloned from mouse brain (Uhler 1993) and rat intestine (Jarchau et al. 1994). Apparently, the type II cGMP kinase gene codes only for a single protein. The human cGMP kinase I and cGMP kinase II genes are located on chromosome 10p11.2-q11.2 (Orstavik et al. 1992) and 4q13.1-q21.1 (Fujii et al. 1995; Orstavik et al. 1996), respectively. Similar to mammals, flies have two genes for cGMP kinase, DG1 and DG2 (Kalderon and Rubin 1989). The *Drosophila* DG1 gene codes for a single protein product, whereas DG2 is transcribed and processed into three major RNA species of different size, T1, T2, and T3 (Kalderon and Rubin 1989).

Sequence comparison of the core region of cGMP kinases, i.e. nucleotide binding and catalytic domain, shows that the bovine cGMP kinase I is more similar to the *Drosophila* genes DG2 and DG1 than to rat cGMP kinase II (Fig. 1B). Furthermore, phylogenetic analyses indicate that cGMP kinase I and DG2 are derived from a common ancestral gene (Fig. 1B) and that the predecessors of cGMP kinase II diverged before the appearance of *Drosophila* DG1 in evolution. Thus, one could speculate that a mammalian branch of DG1 exists, i.e. a not yet identified third form of mammalian cGMP kinase.

3 Structure

3.1 Overall Structure

The cGMP and cAMP kinase belong to the family of serine/threonine kinases that are activated by cyclic nucleotides and share common structural features (Hanks and Hunter 1995). The enzymes are composed of three functional domains, an amino-terminal (A), a regulatory (R) and a catalytic (C) domain (Fig. 1A). The regulatory domain is composed of two in-tandem cyclic nucleotide-binding sites, whereas the catalytic domain contains the MgATP- and peptide-binding pockets. This latter domain catalyses the

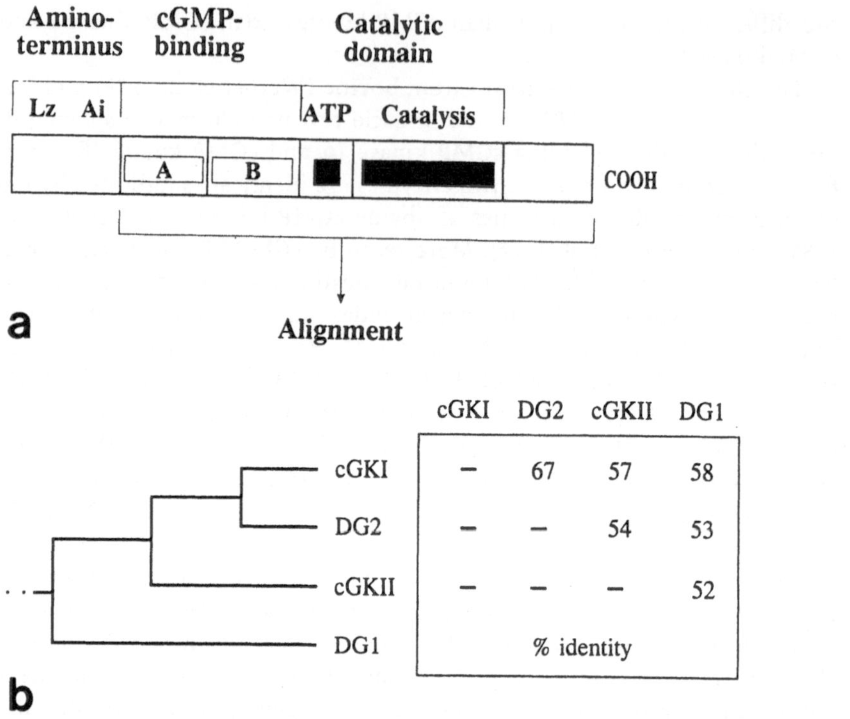

Fig. 1a, b. Structural features of cGMP kinases and phylogenetic analyses. (**a**) Arrangement of functional domains of cGMP kinases. Lz, leucine zipper; Ai, autoinhibitory region; A and B, cGMP binding pockets A and B, respectively. (**b**) Phylogenetic tree of the four mammalian and *Drosophila* cGMP kinases. The tree was calculated on the basis of sequence alignments containing the cGMP-binding pockets, catalytic domains and carboxy-terminus. The numbers indicate the percentage of amino acid sequence identity

transfer of the γ phosphate from ATP to a serine/threonine residue of the target protein.

Mammalian cGMP kinases (Hofmann et al. 1992; Gamm et al. 1995), *Drosophila* DG1 (Foster et al. 1996), and presumably *Drosophila* DG2-T1 and DG2-T3 (Kalderon 1989) are dimers, whereas the enzymes purified from *Paramecium* (Miglietta and Nelson 1988), *Tetrahymena* (Murofushi 1974), and *Dictyostelum discoideum* (Wanner and Wurster 1990) are apparently monomeric enzymes. It is not clear, if this difference in subunit composition is inherent to these enzymes or the result of a proteolytic cleavage of dimeric enzymes (Hofmann et al. 1992), since partial proteolysis of mammalian cGMP kinase can result in a monomeric enzyme that can be activated by cGMP (Wolfe et al. 1989).

Inactive cAMP kinase is a tetramer of two R and two C subunits (R_2C_2) in which the two R subunits tightly bind together. Binding of cAMP dissociates the holoenzyme into a $R_2(cAMP)_4$ dimer and two active C subunits. In contrast, the regulatory and catalytic domain of the cGMP kinase are present on a single polypeptide chain (Fig. 1A). Binding of cGMP does not dissociate the dimeric enzyme into R and C subunits. Native cGMP kinase is an asymmetric molecule with an ellipsoid shape. The binding of cGMP to the enzyme increases the α helical content of the enzyme (Landgraf et al. 1990) and leads to an elongation of the enzyme as indicated by a 23% increase in the axial ratio of the enzyme (Zhao et al. 1997). These conformational changes induce autophosphorylation of the A domain and preceed activation of the enzyme and the phosphate transfer from ATP to the serine or threonine residue of a substrate.

3.2 Amino-Terminus

The amino-terminus of cGMP kinase regulates four important functions: dimerization of homologous subunits, autoinhibition of the catalytic domain in the absence of bound cGMP, the affinity and cooperative behaviour of the cGMP-binding sites A and B, and the intracellular localization of the enzymes.

The dimerization site of cGMP kinase Iα, Iß and II consists of a α-helix with a hydrophobic leucine/isoleucine zipper motif, containing a leucine/isoleucine residue at every first out of seven amino acid residues (i.e. heptad-repeat) (Landgraf et al. 1990; Atkinson et al. 1991). This motif is also found in the *Drosophila* cGMP kinases DG1 and DG2 (Kalderon 1989), but not in the regulatory subunit of the cAMP kinase (Hofmann et al. 1992). cGMP kinase Iα, Iß and II contain 5 to 6 heptad repeats flanked by leucine/isoleucine residues (Fig. 1C). Recombinant expressed cGMP kinase Iα, and Iß are homodimers (Ruth et al. 1991). The proteolytic removal of the amino-terminal domain from the Iα and the Iß isozymes results in an active but monomeric cGMP kinase (Heil et al. 1987; Wolfe et al. 1989) proving the importance of the leucine zipper motif for the dimerization of the subunits. Analysis of the native and the recombinant cGMP kinase II clearly showed that cGMP kinase II is a homodimer (Gamm et al. 1995; Vaandrager et al. 1997). DG1 which contains only three leucine/isoleucine repeats has amino-terminal three valine residues at every seventh position (Fig. 1C). It was postulated that they substitute for leucine/isoleucine in the predicted zipper

(Foster et al. 1996). This interpretation has been confirmed by recent studies on DG1 expressed in Sf9 (*Spodoptera frugiperda*) cells that demonstrated the homodimeric nature of this enzyme (Foster et al. 1996). Sequence similarities suggest that DG2 also is a homodimer (Fig. 1C).

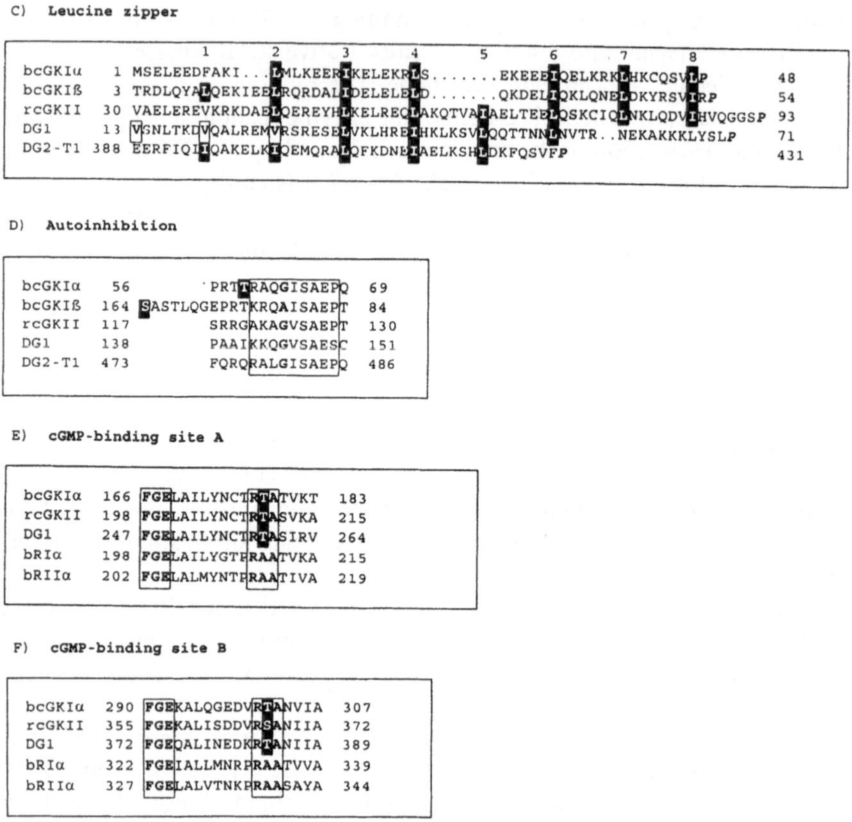

Fig. 1C–F. Structural features of cGMP kinases. Sequence comparison of (**C**) the dimerization site (leucine zipper), (**D**) the autoinhibitory region, (**E**) the cGMP-binding site A and (**F**) the cGMP-binding site B. Leucine zipper: leucines/isoleucines forming the heptad repeat are indicated by black boxes, prolines terminating the leucine containing helices are shown in bold and italic letters; autoinhibitory region: autophosphorylated residues are highlighted in black, the putative autoinhibitory residues are in bold letters; cGMP-binding sites A and B: the conserved Thr and Ser are boxed; bold letters indicate invariant residues. The following sequences were aligned: bovine cGMP kinase Iα (bcGMP kinase Iα), bovine cGMP kinase Iß (bcGKI), rat cGMP kinase II (rcGKII), *Drosophila* cGMP kinases DG1 and DG2-T1, regulatory subunits Iα (bRIα) and IIα (bRIIα) of bovine cAMP kinase

Studies on partially digested cGMP kinase protein revealed the autoinhibitory function of the amino-terminus. Removal of the amino-terminal 78 residues of cGMP kinase Iα, i.e. the leucine zipper and carboxyl-terminal sequences, by tryptic cleavage results in a monomeric, constitutive active kinase which binds two moles of cGMP per mole of enzyme (Heil et al. 1987). In contrast, a cGMP kinase Iß that lacks the leucine zipper (residues 1–62) is a monomeric enzyme, but its activation still depends on cGMP (Wolfe et al. 1989; Smith et al. 1996). These observations imply that the catalytic center is inhibited by the amino-terminus in the absence of bound cGMP and that the autoinhibitory region lies between the dimerization region and the first cGMP binding site. The autoinhibitory region has been identified as a pseudosubstrate sequence (Kemp and Pearson 1991) which is present in each cGMP kinase (Fig. 1D). This sequence has the general structure K/R-K/R-X-G/A-I/V-S-A-E-P/S. In this sequence, glycine or alanine replaces the phosphate accepting serine or threonine.

In the presence of cGMP or cAMP, the mammalian cGMP kinases autophosphorylate serine or threonine residues which are in close proximity to the pseudosubstrate sequence. Autophosphorylation of Thr59 of cGMP kinase Iα (Aitken et al. 1984), Ser64 and Ser80 of cGMP kinase Iß (Smith et al. 1996) and an unidentified amino acid of cGMP kinase II (Vaandrager et al 1997) presumably preceeds activation of the enzymes. Autophosphorylation increases the affinity of cGMP kinase Iα (Hofmann et al. 1985; Landgraf et al. 1986) and cGMP kinase Iß (Smith et al. 1996) for cyclic AMP and enhances the basal activity of the cGMP kinase Iß (Smith et al. 1996). The physiological relevance of cGMP kinase autophosphorylation is not known, but may cause an increased *in vivo* affinity of cGMP Kinase I for cAMP.

Although the cGMP kinase Iα and Iß isozymes differ only in their first 89 and 104 amino-terminal residues, respectively, they are activated at 15-fold different concentrations of cGMP (Ruth et al. 1991). Whereas cGMP kinase Iα exhibits a high and low affinity cGMP binding site with K_d values of 10 and 150 nM, only low affinity sites (K_d 150 nM) are found in cGMP kinase Iß (Ruth et al. 1991). High affinity binding to cGMP kinase Iα is based on positive cooperativity between the two cGMP binding sites, i.e. binding of one molecule cGMP facilitates the binding of another cGMP molecule, and is reflected in a Hill coefficient of >1.5. This cooperativity becomes most obvious by the dissociation of bound cGMP after the addition of excess unlabelled ligand. Under this condition fast and slow (biphasic) dissociation of bound ligand was observed for cGMP kinase Iα, but only fast (monophasic) dissociation for cGMP kinase Iβ, suggesting that the cooperativity is diminished or even abolished in the latter isozyme. Removal of the amino-terminus leads to a loss of cooperativity in cGMP kinase Iα, and therefore it was postulated that positive cooperativity is either achieved by dimerization or by an inherent property of the amino-terminal sequences (Landgraf and Hofmann 1989). Analysis of recombinant chimeric cGMP kinase Iα/Iß proteins identified the amino-terminal residues responsible for cooperativity in cGMP kinase Iα and thereby for high affinity binding of cGMP (Ruth et al. 1997). These residues reside within the leucine zipper and the autoinhibitory domain. A synergistic interaction of both domains is required to develop their regulatory effects on cGMP binding affinity (Ruth et al. 1997). Thus, a new role has been assigned to these two domains, apart from their known function in holoenzyme formation and inhibition of phosphotransferase activity, respectively. As in cGMP kinase Iß, cGMP kinase II has a reduced affinity and cooperativity for the binding of cGMP (Vaandrager et al. 1997; Pöhler et al. 1995). It is anticipated that the largely reduced cooperativity in cGMP kinase II may also be due to residues within the leucine zipper and the autoinhibitory domain. In summary, these observations clearly indicate that the amino-termini have important influence on the concentration of cGMP needed to activate the isozymes, although the amino-termini are not part of the cGMP binding domains.

Another important function of the amino teminal domain is their ability to localize the enzyme to different subcellular structures. cGMP kinase I and II are acetylated and myristoylated at the amino terminus, respectively (Takio et al. 1984; Vaandrager et al. 1996). Myristoylation of cGMP kinase II is required to localize the enzyme to the plasma membrane and to phosphorylate the intestinal chloride channel CFTR (Vaandrager et al. 1998). The transfer of the myristoylation site from cGMP kinase II to Iß allows the phosphorylation of CFTR which does not occur with the native Iß enzyme. A tight association between cGMP kinase and its substrate has been also observed for cGMP kinase Iα and vimentin (MacMillan-Crow and Lincoln 1994). Phosphorylation of vimentin occurs only after dimeric cGMP kinase Iα bound with high affinity in stoichiometric concentrations to vimentin. Neither substrate phosphorylation nor binding occurred after removal of the amino terminal domain. These two examples clearly indicate that the amino-terminal domain is important for the proper organization of the substrate enzyme complex. In contrast to cAMP kinase, anchoring proteins are not required for the correct subcellular localization of cGMP kinase.

3.3 cGMP-Binding Domains

Cyclic nucleotide binding sites of all known subtypes of cAMP/cGMP kinases display a remarkable sequence homology (Fig. 1E and F) (Weber et al. 1989; Shabb and Corbin 1992). This homology extends even further to the cyclic nucleotide-gated ion channels and to the evolutionary more distant *E. coli* Catabolite gene Activator Protein, CAP (Weber et al. 1982). With the crystal structures of two cAMP receptor proteins, CAP (Weber and Steitz 1987) and the R-subunit of cAMP kinase (Su et al. 1995), a detailed picture now emerges of how cyclic nucleotides are recognized by their receptors and what forces govern the ligand binding selectivity. In both proteins, the polypeptide chain of the cAMP-binding domain folds in a similar tertiary motif, consisting of a β-barrel formed by 8 anti-parallel β-sheets (β1–8) and three flanking α-helices. The specificity for cyclic nucleotides, in general, is conserved in these two structures and relates to specific interactions between invariant residues within β6 and β7, and the cAMP ribose and the phosphate diester moieties. However, subtle changes in the secondary structure and unique interactions with the adenine base account for the different cAMP binding affinities seen. Key residues in recognizing cyclic nucleotides are (i) an invariant glutamate within the conserved consensus sequence F-G-E (Fig. 1E and F) which hydrogen bonds to the riboside 2'-hydroxyl group and (ii) an invariant arginine residue which chelates the cyclic phosphate diester as part of the conserved sequence R-T/A-A (Fig. 1E and F). The significance of these residues for binding cyclic nucleotides has been demonstrated by kinase negative mutants (Steinberg et al. 1991), site-directed mutagenesis (Herberg et al. 1996) and cyclic nucleotide analogs (Dostmann and Taylor 1991). A structural model for the cGMP binding sites in the regulatory domain of cGMP kinase also shows the identical conserved residues interacting with the ribose/phosphate region of cGMP (Dostmann, unpublished results). However, there also appear to be distinct protein ligand interactions unique for this receptor site as well as unique for the guanine base. The adenine binding pockets in cAMP kinase consist almost entirely of hydrophobic residues providing a uniquely non-polar environment for adenine. The guanine binding pockets, however, appear to be far more polar providing more opportunities for hydrogen bond interactions between the functional groups of the base and the receptor sites. For instance, Thr178 (Fig. 1E) has been identified as a key residue in interacting specifically with the guanine base in binding site A of cGMP-kinase (Shabb and Corbin, 1992). Thus, the architecture of the cyclic nucleotide binding sites in both enzymes consists of a conserved binding pocket for the ribose-phosphate moiety and a second site that has specificity for either the guanine or adenine base. Table 1 gives an overview on the relative affinity of various cyclic nucleotides for binding sites A and B in cAMP and cGMP kinases.

Table 1. Cyclic-nucleotide activators and inhibitors of cAMP kinase and cGMP kinase isozymes

	cAMP Kinase I	cAMP Kinase II	cGMP Kinase Iα	cGMP Kinase Iβ	cGMP Kinase II
Potent Activators	6-Phe-cAMP [a] 8-Cl-cAMP [b]	6-Phe-cAMP [a] Sp-5,6-DCl-cBIMPS [c]	8-pCPT-cGMP [k] 8-Br-cGMP [l]	8-pCPT-cGMP [k] 8-Br-PET-cGMP [m] 8-Br-cGMP	8-Br-cGMP [l] 8-pCPT-cGMP [k]
Selective Activators	8-Cl-cAMP 6-MB-cAMP [a] 8-AHA-cAMP [d] Sp-8-PIP-cAMPS [c]	Sp-5,6-DCl-cBIMPS [c] 6-MBC-cAMP [a]	8-APT-cGMP [m]	8-Br-PET-cGMP [l]	?
PDE-resistent Activators	Sp-8-Br-cAMPS [e] Sp-8-PIP-cAMPS [e]	Sp-8-Br-cAMPS Sp-5,6-DCl-cBIMPS [e]	8-pCPT-cGMP [k]	8-pCPT-cGMP	8-pCPT-cGMP
Isozyme selective Activators	8-PIP-cAMP 8-AHA-cAMP [f]	6-MBC-cAMP Sp-5,6-DCl-cBIMPS [f]	8-Br-cGMP 1-Me-cGMP [n]	?	?
Inhibitors	Rp-cAMPS [g] Rp-8-Cl-cAMPS [h] Rp-8-Br-cAMPS [j]	Rp-cAMPS [g] Rp-8-pCPT-cAMPS [i]	Rp-8-Br-cGMPS [o] Rp-8-pCPT-cGMPS [p] Rp-8-Br-PET-cGMPS [q]	Rp-8-Br-cGMPS Rp-8-pCPT-cGMPS Rp-8-Br-PET-cGMPS	(Rp-8-Br-cGMPS) Rp-8-pCPT-cGMPS Rp-8-Br-PET-cGMPS

[a] Ogreid et al(1989); [b] Ally et al (1988); [c] Dostmann et al (1989); [d] Skalhegg et al (1992); [e] Schaap et al (1993); [f] Sandberg et al (1991); [g] Dostmann (1995); [h] Yokozaki et al (1992); [i] Weisskopf et al (1994); [j] Gjertsen et al (1995); [k] Geiger et al (1992); [l] Rapoport et al (1982); [m] Sekhar et al (1992); [n] Corbin et al (1986; [o] Zhuo et al. (1994); [p] Gamm et al (1995); [q] Butt et (1995);

Abbreviations: cAMPS, adenosine-3´,5´-cyclic monophosphorothioate; cGMPS, guanosine-3´,5´-cyclic monophosphorothioate; MB, monobutyryl; AHA, aminohexyl; PIP, piperidin; Br, bromo; MBC, mono-t.butylcarbamoyladenosine; APT, aminophenylthio; PET, ß-phenyl-1,N2-etheno; pCPT, chlorophenylthio; Cl, chloro; Phe, phenyl.

Homologies of binding sites. The amino-terminal cGMP binding site A of the different cGMP kinases are more closely related to each other than to the carboxyl-terminal binding site B within the same protein, which holds also true for the cAMP binding sites of the R subunits of cAMP kinases (Shabb and Corbin 1992). Phylogenetic analyses clearly show that the A binding sites of the cGMP and cAMP kinase are more closely related to each other than to the respective B binding sites of the same enzyme.

The biochemical properties of the A and B binding sites differ significantly from each other within one protein kinase. The amino-terminal site A binds cAMP with lower affinity than the carboxy-terminal binding site B, since cAMP dissociates faster from site A than from site B (Rannels and Corbin 1981). cGMP kinase Iα also contains two different cGMP binding sites with fast and slow dissociation behaviour. Based on the sequence homology between the cyclic nucleotide binding sites of cGMP and cAMP kinase (Fig. 1E and F), it was assumed that cGMP dissociates with fast and slow kinetics from the amino-terminal site A and carboxyterminal site B, respectively. However, mutation of the binding sites of cGMP kinase I clearly identified the amino-terminal site (A) as the site with slow dissociation characteristics and a high binding affinity (Reed et al. 1996). This finding clearly demonstrates the limitation of the similarity approach as shown in Fig. 1E and F. It is quite obvious that the used programmes are very limited in respect to the prediction of the functional significance.

3.4 Catalytic Domain

General structure. The eukaryotic protein kinases share a conserved two-lobed catalytic core consisting of a small ATP-binding domain and a large domain that includes the residues that participate in peptide binding and catalysis (Taylor and Radzio-Andzelm 1994; Sicheri et al. 1997). Catalysis takes place in the cleft between the two domains. Most of the residues conserved throughout the protein kinase family cluster around the active site cleft. All protein kinases are only active, if a Thr- and/or Tyr-residue is phosphorylated which is close to the active center (for review see Hanks and Hunter 1995). In many kinases this step is reversible allowing reversible activation and inactivation of the kinase. In contrast, the catalytic center of cAMP kinase and cGMP kinase I is permanantly phosphorylated at Thr197 (Steinberg et al. 1993) and Thr516 (Feil et al. 1995), respectively[1]. The absence of a phosphate group at these positions in the recombinant enzymes

[1] amino acid numbering for cGMP kinase Iα starts with Ser according to Takio et al. 1984.

	ATP-Binding	Linker
bcGKIα	359 FNIIDTLGVGGFGRVELVQLKSEES KTFAM 388	431 KYLYMLMEACLGGELWTILRDRGS 454
rcGKII	453 LEIIATLGVGGFGRVELVKVKNEN I AFAM 481	524 KYVYMLLEACLGGELWSILRDRGS 547
DG1	457 LEVVSTLGIGGFGRVELVKAHHQDRVDIFAL 487	530 KYVYMLLEACMGGEIWTMLRDRGS 553
DG2-T1	777 LRVIATLGVGGFGRVELVQ TNGDSSRSFAL 806	849 KYLYMLMESCLGGELWTILRDKGN 872
rcAKCα	43 FDRIKTLGTGSFGRVMLVKHKESG NHYAM 71	114 SNLYMVMEYVPGGEMFSHLRRIGR 137

	Catalytic loop	Activation loop	Boundary Catalytic/-COOH
bcGKIα	481 YRDLKPEN 488	511 GKKTWT 516	601 RLGNLKNGVKDIQKHKWF 618
rcGKII	574 YRDLKPEN 581	604 GQKTWT 609	694 RLGNLKNGINDIKKHRWL 711
DG1	580 YRDLKPEN 587	610 SSKTWT 615	700 RLGYQTGGIQDIKKHKWF 717
DG2-T1	899 YRDLKPEN 906	929 GRKTWT 934	1019 RLGYQRGGISEIQKHKWF 1036
rcAKCα	164 YRDLKPEN 171	193 GR TWT 197	280 RFGNLKNGVNDIKNHKWF 297

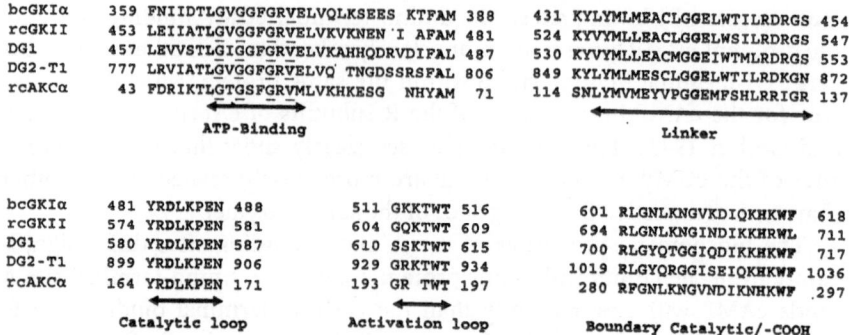

Fig. 2. Sequences participating in catalysis. Sequence comparison of the ATP-binding region, the linker between the two lobes of the catalytic domain, the catalytic loop, the activation loop and the boundary between catalytic domain and the carboxyterminal part of the kinases. Underlined amino acid residues represent the consensus motif for ATP-binding. Bold letters indicate (i) the residues forming the pocket for the adenine ring (ATP-binding), (ii) an invariant glycine interacting with the substrate (linker), (iii) the catalytic base (catalytic loop), (iv) the phosphorylated threonine (activation loop), and (v) invariant residues in the carboxyterminal part of the catalytic domain. Following sequences were aligned: bovine cGMP kinase Iα (bcGK Iα), rat cGMP kinase II (rcGKII), *Drosophila* cGMP kinases DG1 and DG2-T1, catalytic subunit of rat cAMP kinase (rcAKCα)

results in catalytically inactive enzymes. These phosphothreonines are located in the so-called activation loop (Fig. 2). This loop seems to be a critical part of the kinase molecule, although it is not identical with the phosphotransfer site. The residues within the activation loop (Gly511 to Thr515) (Fig. 2) stabilize the subdomain in a conformation permitting proper orientation towards the substrate peptide by forming hydrogen bonds between the phosphate oxygens of phospho-Thr516 with charged side chains of other residues.

Catalysis. In analogy to the crystal structure of the ternary complex consisting of cAMP kinase catalytic subunit Cα, peptide inhibitor (PKI 5-24) and MgATP (Zheng et al. 1993), several residues in the catalytic domain of cGMP kinases are thought to play an essential role in catalysis: The catalytic domain contains at the amino-terminus the consensus motif Gly-X-Gly-X-X-Gly-X-Val (starting with Gly366) for ATP binding (Fig. 2). It has been postulated that the invariant Leu365, Val373 and Ala387 in cGMP kinase contribute to a hydrophobic pocket that encloses the adenine ring of ATP. The boundary between the catalytic domain and cGMP binding site B is located seven residues amino-terminal to the first glycine, where a hydrophobic residue (Phe359) is usually found (Hanks and Hunter 1995). A stretch of amino acid residues (Lys431–Ser454) links the two lobes of the catalytic domain. The invariant Glu443 herein is thought to participate in peptide binding by forming an ion pair with the first basic residue (Arg/Lys) in the substrate consensus recognition site Arg/Lys-Arg/Lys-X-Ser/Thr. The motif Tyr481-Arg-Asp-Leu-Lys-Pro-Glu-Asn488 has been termed the catalytic loop because Asp483 within this loop has emerged as the candidate for the catalytic base accepting the proton from the attacking substrate hydroxyl group during the phosphotransfer reaction. Lys485 in the loop may help to facilitate the phosphotransfer by neutralizing the negative charge of the γ-phosphate during transfer. A notable feature of the carboxyl-terminal part of the catalytic domain is Arg601, which is conserved in most kinases. In many serine/threonine kinases, the motif His-X-aromatic-hydrophobic amino acid is found about 10 residues downstream of the invariant Arg and appears to define the boundary between the catalytic domain and the carboxyl-terminal domain. Like in cAMP kinase (Chestukhin et al. 1996), the carboxyl-terminal tail of the cGMP kinase is assumed to contribute to substrate recognition.

Peptide substrate requirements. Various synthetic peptides originally derived from *in vitro* substrate proteins were used initially to analyse the requirements for a peptide substrate. Thereafter, optimum sequences were obtained by an iterative library screen of decameric peptides (Tegge et al. 1995). As summarized in Table 2A, a possible peptide sequence is TQAKRKKSLA(MA) which combines a micromolar K_m value with an acceptable velocity. These peptides are characterized by only a 2- to 6-fold preference for GMP kinase versus cAMP kinase. A higher selectivity for cGMP kinase I is usually obtained only at higher K_m and/or lower V_{max} values. cGMP kinase I and cGMP kinase II differ in their substrate selectivity (Table 2B). Several substrates have been investigated recently, however, only the peptide derived from the cAMP response element binding protein (CREB) exhibits some selectivity towards cGMP kinase II (Gamm et al. 1995), whereas the histone 2B substrate, which is readily phosphorylated by cGMP kinase I, is a very poor substrate of cGMP kinase type II. This suggests that both cGMP kinase isoforms interact differently with substrates, probably due to amino acid differences within the catalytic domains. In some peptides, a phenylalanine carboxy-teminal to the phosphate accepting serine/threonine decreases the affinity of the peptide for cAMP kinase but not that for cGMP kinase I. Thus, at least *in vitro*, the surroundings of the phosphate acceptor site in the substrate peptide contributes significantly to the efficacy and selectivity of its phosphorylation by cGMP and cAMP kinase (Colbran et al. 1992).

Table 2A. Peptide substrates of cAMP kinase and cGMP kinase Iα

Substrate		cAMP kinase		cGMP kinase Iα		Specificity
		K_m (μM)	V_{max} μmol/min/mg	K_m (μM)	V_{max} μmol/min/mg	(cGK/cAK)
$[S^{21}]$-$PKI^{(6-22)}$ [a]	TTYADFIASGRTGRRNSIHD – NH_2	0.3		12.0		0.03
$[S^{21}]$-$PKI^{(14-22)}$ [a,b]	GRTGRRNSI-NH_2	0.1		2.3	11.0	
Kemptide [b]	LRRASLG	4.3	9.9	120.0	4.5	0.02
4-3 [c]	RAERRASI	2.0	8.0	14.0	2.6	0.05
$[S^{32}]$-$H2B^{(29-35)}$ [d]	RKRSRKE	113.0	1.2	21.6	4.4	20
VASP [e]	LRKVSKQE	1395.0	2.6	94.0	3.7	21
VASP [e]	IERRVSNAG	26.0	2.7	30.0	2.2	0.7
Aminoalcohol[f]	LRRRRF-NH-$(CH_2)_3$OH	54.0	9.9	0.6	3.6	33
WW4 [c]	TQAKRKKSLA-NH_2	2.7	8.0	1.7	11.1	2.2
WW15 [g]	TQAKRKKSLAMA-NH_2	2.5	8.0	0.6	10.9	6.2

Table 2B. Peptide substrates preferentially phosphorylated by cGMP kinase II

Substrate		cGMP kinase II		cGMP kinase Iα		Specificity
		K_m (μM)	V'_{max} μmol/min/mg	K_m (μM)	V_{max} μmol/min/mg	(cGKII/cGKI)
CREBtide [h]	KRREILSRRPSYR	16	1.5	37	1.5	2.3
IP_3Rtide [h]	GRRESLTSFG	10	1.1	48	4.8	1.1

[a] Mitchell et al (1995); [b] Hofmann et al. (1992); [c] Tegge et al. (1995); [d] Glass and Krebs (1979); [e] Butt et al. (1994); [f] Wood et al (1996); [g] Dostmann et al. (unpublished results); [h] Gamm et al. (1995).

Protein kinase inhibitor peptides. Replacement of serine/threonine residues, which are phosphorylated in the substrate peptides by alanine may result in an inhibitory peptide (Table 3). Among these peptides, the WW7 and WW21 (see Table 3) derived from the "optimal“ substrate peptide KRKKSL (Tegge et al. 1995) exhibit high selectivity towards cGMP kinase I. A classical example of an inhibitor is the native heat-stable protein kinase inhibitor (PKI) that interacts with cAMP kinase with an K_i of 2 nM. PKI and peptides derived from it such as PKI(5-24) block the phosphotransferase activity of cAMP kinase through the pseudosubstrate sequence RRNA. The crystal structure of the ternary complex showed that the aromatic residues Tyr235 and Phe239, in concert with Pro236, form a sandwich-like structure which interacts with Phe10 of PKI(5-24) (Knighton et al. 1991). PKI(5-24) exhibits a significant degree of selectivity, since 5 orders of magnitude higher concentrations (apparent K_i 150 μM) are needed to block cGMP kinase I as compared with cAMP kinase (Glass et al. 1986) (Table 3). A possible explanation might be the fact, that the critical Tyr235 and Phe239 of cAMP kinase are replaced by two serines in cGMP kinase I. However, these residues are likely not the sole cause of this difference between cGMP and cAMP kinase. Replacement of Ser555 and Ser559 by Tyr and Phe in cGMP kinase I, which should provide the hydrophobic pocket needed for high affinity binding of PKI, reduced only slightly the K_i of the mutant cGMP kinase I for PKI to 40 μM (Ruth et al. 1996).

A key difference between the catalytic centers of the cGMP and cAMP kinases is the Ser53 residue in the phosphate anchor region (P-loop) of cAMP kinase which interacts with the γ-phosphate of ATP and the substrate backbone (Bossemeyer et al. 1993). The equivalent residue in cGMP kinase is a glycine which does not provide the same interactions. Arg133 of cAMP kinase has been suggested to increase the hydrophobicity of the pocket for Phe10 of PKI (Wen and Taylor 1994). This residue is conserved in cGMP kinase. The subsequent amino acid Arg134 in cAMP kinase, however, is replaced by an Asp in cGMP kinase I, suggesting that a negative charge close to the hydrophobic pocket may contribute to the extremely low affinity for PKI. These and other changes probably cause the molecular basis for the different interaction of cGMP kinase with substrate peptides and may contribute to the fact that cGMP kinase and cAMP kinase phosphorylate distinct substrates *in vivo*.

Inhibitors of ATP binding. A large family of isoquinoline sulfonamide compounds (Hidaka and Kobayashi 1992) and semisynthetic derivatives of compounds isolated from *Nocardiopsis* (Kase et al. 1987) inhibit protein kinases by competing with ATP binding. Among these compounds are the H- and

Table 3. Peptide inhibitors of cAMP kinase and cGMP kinase Iα

Inhibitor		cAMP kinase K_i (μM)	cGMP kinase Iα K_i (μM)	Specificity (cAK/cGK)
$[A^{21}]$-$PKI^{(5\text{-}24)}$ [a,b]	TTYADFIASGRTGRRNAIHD – NH_2	0.002	150	0.00001
$[A^{21}]$-$PKI^{(14\text{-}22)}$ [a]	GRTGRRNAI-NH_2	0.07	47	0.001
Ala-Kemptide [c]	LRRAALG	800	350	2.3
$[A^{32}]$-$H2B^{(29\text{-}35)}$ [d]	RKRARKE	550	86	6.4
	LRRRRFAFC(Npys) [e]	25	15	1.7
	LRRRR$F_{(D)}$AFC(Npys) [e]	no inactivation	21	∞
WW7 [f]	TQAKRKKALA-NH_2	685	15	46
WW21 [f]	TQAKRKKALAMA-NH_2	750	8	107

[a] Glass et al. (1992); [b] Ruth et al (1996); [c] Glass (1983). [d] Glass and Krebs (1982); [e] Yan et al. (1996); [f] Dostmann et al. (unpublished results).

the K-252-series of kinase inhibitors which block catalytic activity of cGMP kinase and cAMP kinase in the submicromolar range. The compounds KT-5823 and KT-5720 are preferentially used to block cGMP kinase and cAMP kinase, respectively (Kase et al. 1987). KT-5823 (K_i 0.25 µM) has a 20–50-fold higher affinity towards cGMP kinase I than to cAMP kinase, PKC, MLCK, and CaMK (Hidaka and Kobayashi 1992). The isoquinoline compounds and the KT-family competitively interfere with the binding of ATP but neither with that of the catalytic domain nor with the ATP-binding site of other enzymes. This selectivity is obtained by interactions of the isoquinoline ring with the kinase specific adenosine pocket (Engh et al. 1996; Xu et al. 1996). The inhibitory mechanism of K-252 compounds appears to resemble that of the isoquinolines. The K-252 compounds exhibit at least one order of magnitude higher inhibitory activity than the isoquinolines.

In vivo substrates. As shown in Table 4, the KRKK motif described for peptide substrates *in vitro* is not essential for *in vivo* substrate proteins. Substrates phosphorylated by cGMP kinase include ligand-operated ion channels such as the IP_3 receptor (Komalavilas and Lincoln 1994, 1996), the ryanodine receptor (Suko et al. 1993) and the $GABA_A$ receptor β_1, β_2, β_3 and β_4 subunits (McDonald and Moss 1994, 1997). However, neither of these sub-

Table 4. Identified phospho-acceptor sites of known cGMP kinase I substrates

VASP	...LRKVSKQEEA.	*in vivo*	Butt 1994
	.HIERRVSNAG...		
G-substrate	...RRKDTPALHI...	*in vitro*	Aitken 1981
	...RRKDTPALHT...		
Histone 2B	....RKRSRKE...	*in vitro*	Hashimoto 1976
Phosphorylase b kinase α subunit	...FRRLSISTE...	*in vitro*	Yeaman 1977
cGMP-binding PDE (PDE V)	....RKISASEFDRPLR.	*in vitro*	Thomas 1990
Rap1B	...ARKKSC.. .	*in vitro*	Miura 1992
IP_3 recepetor	...GRRESLTSFG...	*in vitro* *in vivo*	Komalavilas 1994 Komalavilas 1996
Ryanodine receptor	.....KISQTAQTYDPR	*in vitro*	Suko 1993
$GABA_A$ receptor ß2 subunit	...RRRASLQK...	*in vitro*	McDonald 1997
$GABA_A$ receptor ß3 subunit	...RRRSSLQK...	*in vitro*	McDonald 1997

strates exhibit selective preference for cGMP kinase, e.g. the synthetic peptide corresponding to serine 1755 (GRRESLTSFG) of the smooth muscle IP_3 receptor was phosphorylated equally well by cAMP kinase and cGMP kinase type Iα with a K_m in the range of 30–40 μM and a V_{max} of 6 μmol/min/mg. A similar potency of cAMP kinase and cGMP kinase I has also been observed for phosphorylation of the skeletal muscle ryanodine receptor at Ser2843 (Table 4.). Apart from cGMP kinase I, several other kinases phosphorylate a conserved Ser residue in $GABA_A$ receptor β subunits (Ser409 in β_1). A ten times stronger selectivity for cGMP kinase I versus cAMP kinase is present in the phosporylation site within the cGMP-specific phosphodiesterase (PDE V) (Thomas et al. 1990). Binding of cGMP to both allosteric sites is required for phosphorylation of the PDE V (Turko et al. 1998) suggesting that the enzyme is a physiological substrate for cGMP kinase. The cerebellar G substrate, a putative phophatase regulator, has a lower K_m for cGMP kinase I than for cAMP kinase (Aitken et al. 1981). In summary, these observations confirm the hypothesis that kinase specificity is provided by co-localization of the cGMP kinase rather than by an optimal recognition motif within the substrate (MacMillan-Crow and Lincoln 1994; Vaandrager et al. 1998).

4
Expression

Expression of cGMP kinases in Drosophila. DG1 and DG2 are already expressed during *Drosophila* embryonic development (Kalderon 1989). Embryonic DG1 expression is temporally restricted to stage 13 embryos and is confined to the cephalic region and to the amnioserosa (Foster et al. 1996). At the cellular level, only a small population of cells have been found to contain DG1 kinase. These cells are characterized as hemocytes and macrophages. In the adult fly, DG1 transcript was found primarily in head tissue (Kalderon 1989), such as optic lobes and proximal cortex (Foster et al. 1996). Only weak DG1-immunoreactivity was observed in the body confined to the testis. DG2 expression has been studied by Northern blot analyses indicating that the major transcripts of DG2 kinase are expressed in adult head and body tissues of the fly (Kalderon 1989).

Tissue distribution of cGMP kinase I and II. Mammalian cGMP kinase has been identified in many different organs and cells of various species. The highest concentrations of cGMP kinase I are found in smooth muscle, platelets and cerebellum (Lohmann et al. 1981; Waldmann et al. 1986; Keilbach et al. 1992). In addition, high levels of cGMP kinase I have been detected in hippocampus (Kleppisch et al. 1998), dorsal root ganglia (Qian et al. 1996), neuromuscular endplate (Chao et al. 1997) and cells of the kidney vasculature (Joyce et al. 1986). Lower levels of cGMP kinase I are expressed in vascular endothelial cells (MacMillan-Crow et al. 1994; Diwan et al 1994; Draijer et al. 1995) and immune cells (Pryzwanski et al. 1995). Immunoblotting of bovine and rat tissues with isozyme-specific antibodies showed that cGMP kinase Iß is highly expressed in uterus, aorta, and trachea but not in lung, heart or cerebellum (Keilbach et al. 1992). The latter tissues contain relatively high concentrations of cGMP kinase Iα. In bone, cGMP kinase I has been found in growth plate chondrocytes (Pfeifer et al. 1996) and osteoclasts (Van Epps-Fung et al. 1994).

cGMP kinase II has been detected only in few tissues so far. Brain regions with high cGMP kinase II expression are thalamus, cortex, olfactory bulb and brainstem (El-Husseini et al. 1995). With the exception of brain and bone, cGMP kinase II is predominantly localized to epithelial cells with secretory properties. cGMP kinase II is present in the brush border of the small intestine (Markert et al. 1995), the proximal convoluted tubules of kidney (Gambaryan et al. 1996), the ciliary epithelium of the epidydimis and the lung (Pfeifer, unpublished results). Within the intestinal mucosa, cGMP kinase II is predominantly expressed in small intestine and proximal colon, whereas only marginal amounts of cGMP kinase II protein are detected in distal colon, caecum and stomach. So far, growth plate chondrocytes are the only cells that have been shown to express cGMP kinase I as well as cGMP kinase II (Pfeifer et al. 1996). cGMP kinase II is expressed mainly at the border between proliferative and hypertrophic chondrocytes, and also – at lower level – in early proliferative, resting and articular chondrocytes, whereas cGMP kinase I is confined to the hypertrophic cells.

Subcellular localization. Native and recombinant cGMP kinase I is mainly a cytosolic enzyme but also has been detected in detergent-solubilized fractions of various tissue (Keilbach et al. 1992) indicating that the enzyme associates with cellular membranes and/or the cytoskeleton. cGMP kinase Iß contains a glycine residue at the second amino-terminal position that might serve as an acceptor for myristoyl groups (Casey 1995), whereas cGMP kinase Iα is acetylated (Takio et al. 1984). By use of confocal laser scanning microscopy it was possible to visualize the cellular distribution of cGMP kinase I in smooth muscle cells, neutrophils and mononuclear phagocytes (Lincoln et al. 1993; Pryzwansky et al. 1995). In isolated smooth muscle cells, cGMP kinase I is most abundant in the cytosol, and very little cGMP kinase I-specific staining was observed at the plasma membrane (Lincoln et al. 1993). Most of the kinase of smooth muscle cells is localized in the perinuclear area which is rich in intermediate filaments and sarcoplasmic reticulum. In neutrophils, cGMP kinase transiently co-localizes with and phosphorylates the intermediate filament protein vimentin after stimulation with formyl-peptide (Wyatt et al. 1991). Interestingly, phosphorylation of vimentin does not occur in non-activated cells treated with cGMP analogues. A second interesting feature of cGMP kinase I in immune cells is its localization at the margins of cells that are actively involved in cell spreading (Pryzwansky et al. 1995) suggesting that cGMP kinase I may be involved in cell adhesion and motility. This hypothesis is supported by the finding that the vasodilator-stimulated phosphoprotein (VASP) is a major substrate for cGMP kinase in platelets and vascular endothelium (Halbrügge et al. 1990; Draijer et al. 1995). VASP is associated with microfilaments and focal adhesion points (Reinhard et al. 1992), and most of cGMP kinase I is localized within the particulate fraction of platelet extracts (El-Daher et al. 1996). Taken together, these observations indicate that at least a fraction of cGMP kinase I is targeted to the cytoskeleton and organelles including the nucleus. A nuclear localization of cGMP kinase is a prerequisit for the postulated direct regulation of gene expression by the kinase (Gudi et al. 1996). Native and recombinant cGMP kinase II resides in the particulate fraction of tissue extracts (Pöhler et al. 1995; Vaandrager et al. 1996; Vaandrager et al. 1998). Membrane localization requires myristoylation of Gly-2 (Vaandrager et al. 1996). Mutation of Gly-2 to Ala (Vaandrager et al. 1996) or the addition of a His-tag to the amino-terminus (Gamm et al. 1995) results in a mostly soluble enzyme.

5
Function

The widespread expression of cGMP kinases is mirrored by the diversity of their functions that establish these enzymes as major mediators of the cGMP signaling cascade. The majority of the functional studies on cGMP kinases over the past few years has been centered on the role these kinases play in smooth muscle, platelets and intestinal epithelium. More recent studies indicate that cGMP kinases play also a pivotal role in kidney, bone and nervous system.

Role of cGMP kinases in the regulation of smooth muscle tone and blood pressure

Smooth muscle tone and blood pressure are regulated by a variety of hormones and other factors including NO and natriuretic peptide. NO and natriuretic peptide decrease vascular smooth muscle tone by stimulation of the cGMP production. Studies in isolated tracheal and vascular smooth muscle strips and cells (Felbel et al. 1988; Francis et al. 1988; Cornwell and Lincoln 1989) established a causal link between activation of cGMP kinase I and relaxation of smooth muscle. Activated cGMP kinase I lowers cytosolic Ca^{2+}concentrations ($[Ca^{2+}]_i$) in smooth muscle and various other cell types (Felbel et al. 1988; Cornwell and Lincoln, 1989; Geiger et al. 1992; Ruth et al. 1993). This effect of cGMP kinase I is restricted to agonist-induced increases of $[Ca^{2+}]_i$, whereas basal $[Ca^{2+}]_i$ is not affected (Ruth et al. 1993; Pfeifer et al. 1998). The available data suggest that cGMP kinase I relaxes smooth muscle by at least two mechanisms:

(i) Inhibition of calcium release from the sarcoplasmic reticulum (SR) either by phosphorylation of the inositol 1,4,5-trisphosphate (IP_3) receptor (Lincoln et al. 1995) and/or by inhibition of agonist-induced generation of IP_3 (Rapoport 1989; Lang and Lewis 1989). NO/cGMP has been shown to inhibit phosphatidylinositol breakdown in rat and rabbit aorta. cGMP kinase I which had been transfected into CHO cells inhibits thrombin-induced IP_3 synthesis and the release of calcium from the SR (Ruth et al. 1993). The target of cGMP kinase I, i.e. hormone receptor, G-protein or phospholipase C, is not clear.
(ii) Inhibition of calcium influx via voltage-dependent calcium channels. Smooth muscle contraction is initiated by calcium release followed by a calcium influx through voltage-dependent L-type calcium channels. Hyperpolarization of the plasma membrane decreases the open probability of the calcium channels and thereby causes a reduction in smooth muscle tension. Ca^{2+} -activated maxi-K^+ (BK_{Ca}) channels significantly contribute to the membrane potential in vascular and bronchial smooth muscle. The open probability of the BK_{Ca} channel is increased after phosphorylation by cAMP kinase (Kume et al. 1989; Kume at al. 1994). Since this process is followed by hyperpolarization of the membrane, it was postulated that through this mechanism, at least in part, the relaxing effects of ß-adrenergic agonists could be explained (Hofmann et al. 1994). The same large conductance BK_{Ca} channel is regulated by cGMP kinase in vascular smooth muscle cells (Taniguchi et al. 1993; Robertson et al. 1993) and mesangial cells (Sansom et al 1997). An indirect regulation of this channel by cGMP kinase through activation of protein phosphatase 2A was identified in neurohypophyseal cells (White et al. 1993), CHO cells, and bovine tracheal smooth muscle (Zhou et al. 1996). Recent experiment with the expressed BK_{Ca} from human myometrium favor a direct regulation of the channel by cGMP kinase I (Alioua et al 1998). Regardless of the mechanism, cGMP kinase I activation resulted in a BK_{Ca} channel which is activated already at lower $[Ca^{2+}]_i$ and more negative

potentials than the non-modified channel. The resulting hyperpolarization impairs Ca^{2+} influx via voltage-dependent Ca^{2+} channels and relaxes the smooth muscle. Earlier experiments indicated the possibility that cGMP kinase I lowers the Ca^{2+} sensitivity of the contractile apparatus (Pfitzer et al. 1984). Together, these two mechansims, i.e. inhibition of $[Ca^{2+}]_i$ transients and decreased Ca^{2+} sensitivity of the contractile elements, could operate in concert with each other, thereby effectively blocking agonist-induced contractions.

A further mechanism which may contribute to the decrease in blood pressure could be the inhibition of renin release by cGMP kinase II. cGMP kinase II has been detected in juxtaglomerular (JG) apparatus of the kidney (Gambaryan et al. 1996). Several in vitro studies in isolated JG cells showed that NO either inhibits (Kurtz et al. 1986; Henrich et al. 1988; Greenberg et al. 1995) or stimulates (Gardes et al. 1992; Scholz and Kurtz 1993) renin secretion. Studies on the isolated perfused JG apparatus indicated that macula densa-generated NO is important for the stimulation of renin secretion whereas endothelial-derived NO inhibits renin secretion (He et al. 1995). Expression of renin transcripts and cGMP kinase II apparently correlate with each other (Gambaryan et al. 1996). In support of such a mechanism, synthesis of renin transcripts and renin secretion is increased in cGMP kinase II deficient mice, but not affected in cGMP kinase I-deficient mice (Wagner et al 1998). In addition, dehydration induces an up-regulation of cGMP kinase II mRNA in the rat papilla and the brush border of the proximal tubule (Gambaryan et al. 1996). Together, these experiments suggest a direct negative regulatory link between cGMP kinase II and renin synthesis. The regulation of renin secretion seems to be not the only function of cGMP kinase II in the kidney. The dehydration-induced increase in cGMP kinase II transcript (Gambaryan et al. 1996) is paralleled by an increase in the expression of a kidney-specific chloride channel ClC-K1 (Uchida et al. 1993). Therefore, it was speculated that – in analogy to the situation in the intestine where cGMP kinase II regulates CFTR Cl- channels – cGMP kinase II may also be involved in ClC-K1 regulation (Gambaryan et al. 1996).

The biological significance of cGMP kinases was ascertained further in mice that lacked both cGMP kinase I isoforms (Pfeifer et al. 1998) and cGMP kinase II (Pfeifer et al. 1996). cGMP kinase I-deficient mice are hypertensive. Their mean arterial blood pressure (MAP) is ~ 20 mm Hg higher than that of wild type (Pfeifer et al. 1998). Interestingly, mice lacking endothelial NO synthase (eNOS) (Huang et al. 1995; Shesely et al. 1996) exhibit a similar increase in MAP. Aortic rings from both cGMP kinase I- and eNOS-deficient mice are refractory to acetylcholine-induced vasodilation (Fig. 3A). As ex-

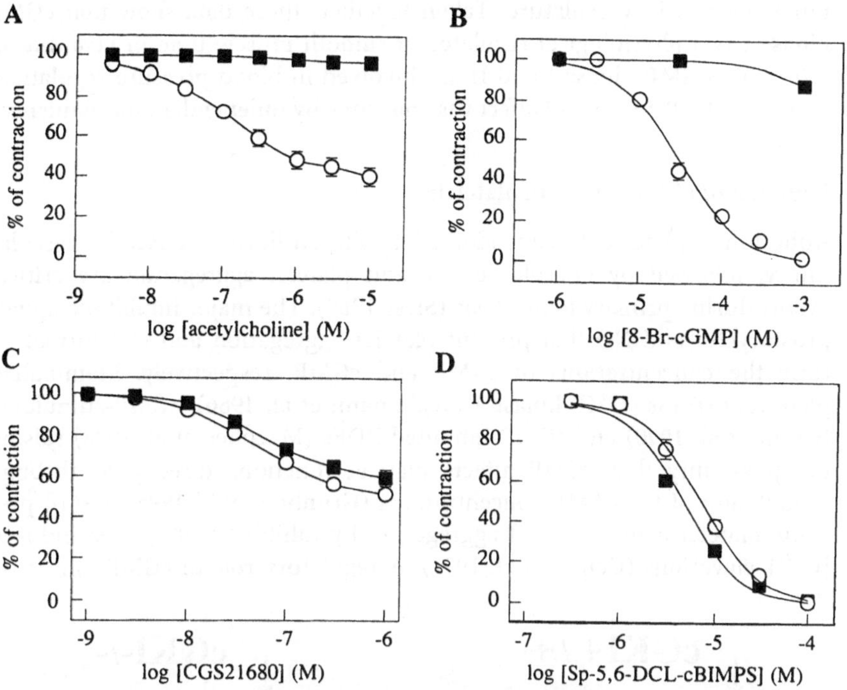

Fig. 3A–D. Effects of acetylcholine (A), 8Br-cGMP (B), adenosine A_2 receptor agonist CGS21680 (C) and Sp-5,6-DCL-cBIMPS (a cAMP analog) (D) on isolated aortic rings with intact endothelium. Aortic rings were from wild type (O) and cGMP kinase I-deficient (■) mice. Values are the mean ± SEM for 12 aortic strips from 3–5 different animals of each genotype. Curves were fitted using a logarithmic function

pected, aortic rings from cGMP kinase I-deficient mice were not relaxed by the cGMP analogue 8-Br-cGMP (Fig. 3B).

Previously, it has been emphasized that in vascular smooth muscle cGMP kinase I is cross-activated by cAMP (Lincoln et al. 1990; Jiang et al. 1992). In contrast to expectation, superfusion of aortic rings from cGMP kinase I-deficient mice with CGS-2160 – an adenosine A_2 receptor agonist that stimulates endogenous cAMP production – relaxed cGMP kinase I-deficient and wild type aortas with EC_{50} values of 60 ± 2 nM and 56 ± 6 nM, respectively (Pfeifer et al. 1998) (Fig. 3C). In addition, no significant difference between the dose-response curves for the two mouse strains was observed when Sp-5,6-DCL-cBIMPS – a potent cAMP analogue – was used (Fig. 3D). These data show that cAMP signalling is not impaired in the absence of cGMP kinase I and suggest that cross-activation of cGMP kinase I is of minor relevance in vasculature. Taken together, these data show that cGMP kinase I is a physiological regulator of smooth muscle tone *in vivo and in vitro*, that cGMP kinase I and II are involved in blood pressure regulation, and that cAMP and cGMP affect vascular tone by independent mechanisms.

Function of cGMP kinase in platelets

Adhesion of platelets to subendothelial collagen fibrils exposed by vascular injury, followed by granule release and platelet aggregation are critical events during primary hemostasis (Siess 1989). The major inhibitory signals prostacyclin and NO that prevent platelet aggregation and clot formation raise the concentrations of cAMP and cGMP, respectively. Mammalian platelets express cGMP kinase I (Waldmann et al. 1986), cGMP-stimulated (Grant et al. 1990) and cGMP-inhibited PDEs (Macphee et al. 1988) raising the possibility that cGMP affects platelet functions directly or through modulation of the cAMP concentrations (Grunberg et al 1995). cGMP prevents platelet activation and aggregation by inhibition of agonist-induced $[Ca^{2+}]_i$ elevations (Geiger et al. 1992). A regulatory role of cGMP kinase on

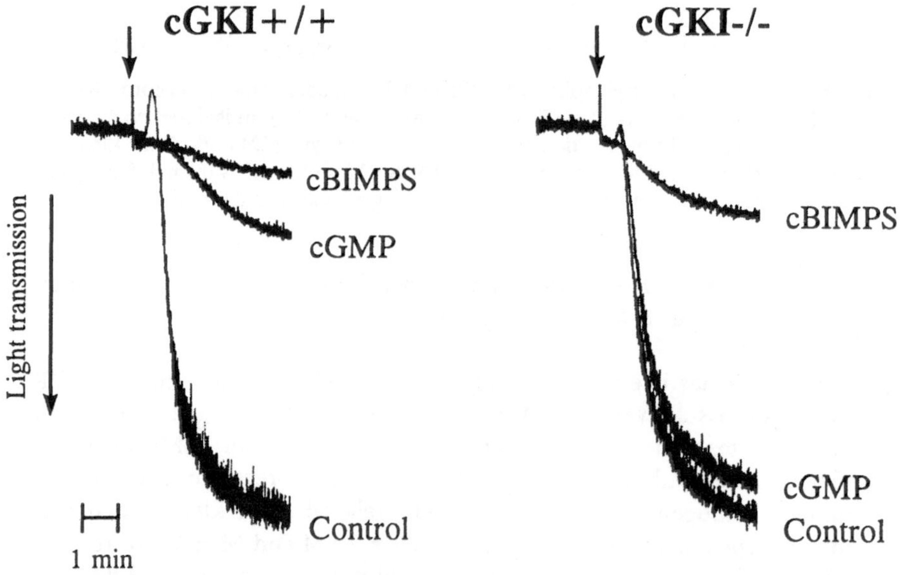

Fig. 4. Aggregation of platelet from wild type (cGKI+/+) and cGMP kinase I-deficient (cGKI-/-) mice. Suspensions of washed platelet preparations were incubated for 20 min with 0.1 mM of the cAMP analog Sp-5,6-DCL-cBIMPS (cBIMPS), 0.1 mM 8-pCPT-cGMP (cGMP) or with 0.1% DMSO (Control). The arrow indicates the addition of collagen (5 µg/ml)

platelet $[Ca^{2+}]_i$ was supported by the finding that cGMP did not prevent agonist induced $[Ca^{2+}]_i$ transients in cGMP kinase-deficient human platelets isolated from chronic myelocytic leukemia (CML) patients (Eigenthaler et al. 1993). The direct involvement of cGMP kinase I was unequivocally proven by the use of platelets from cGMP kinase I-deficient mice. The addition of 5 µg/ml collagen activates and aggregates wild type as well as mutant platelets. Preincubation of wild type platelets with 8pCPT-cGMP, a potent activator of cGMP kinase, inhibited shape change by 93%, aggregation by 68% and secretion of serotonin by 85% (Fig. 4). In contrast, 8pCPT-cGMP did not affect these parameters in cGMP kinase I -deficient platelets nor did it induce phosphorylation of the focal adhesion protein VASP, a major substrate for both cGMP and cAMP kinase *in vitro* and *in vivo* (Butt et al. 1994). A recent report suggests that cGMP kinase I phosphorylates the platelet thromboxane receptor and inhibits thereby signalling through the Gq-phospholipase C-IP_3 pathway (Wang et al. 1998). As already observed with the vascular smooth muscle, the signaling of cAMP was not affected in the cGMP kinase I-deficient platelets showing again that cGMP and cAMP affect cellular functions by independent pathways.

Role of cGMP kinases in the regulation of intestinal functions

The role of NO as a major regulator of smooth muscle tone is not confined to the cardio-vascular system but extents to the gastro-intestinal tract. Inhibitory nonadrenergic-noncholinergic (NANC) neurons that reside in the intestinal wall have been shown to contain NOS and to release NO (Desai et al. 1991; Burns et al. 1996). The NANC motoneurons, together with interstitial cells of Cajal (ICC) and smooth muscle cells, compose the basic motoric unit of the intestine, which is essential for the accommodation of food and generation of peristaltic waves. NO released from NANC neurons stimulates cGMP production and relaxes smooth muscle. Several mouse models have been described with developmental defects in inhibitory NANC neurons (Greenstein-Baynash et al. 1994; Hosoda et al. 1994; Shirasawa et al. 1997) or cells of Cajal (Maeda et al. 1992; Huizinga et al. 1995; Burns et al. 1996) that exhibit severe intestinal distension.

cGMP kinase I is highly epxressed in the intestinal smooth muscle (Fig. 5C, E). Autopsy of cGMP kinase I-deficient mice revealed gross distension of the gastrointestinal tract with clear signs of pyloric stenosis (Fig. 5A, B) and a marked hypertrophy of the muscle layer (Fig. 5D, F). Pyloric stenosis is also a characteristic of mice lacking a splice variant of the nNOS gene ($nNOS^{D/D}$) (Huang et al. 1993; Eliasson et al. 1997) and of humans with a decreased NOS activity (Vanderwinden et al. 1992). The NOS and vasoactive intestinal polypeptide (VIP)-immunoreactivity was not different between

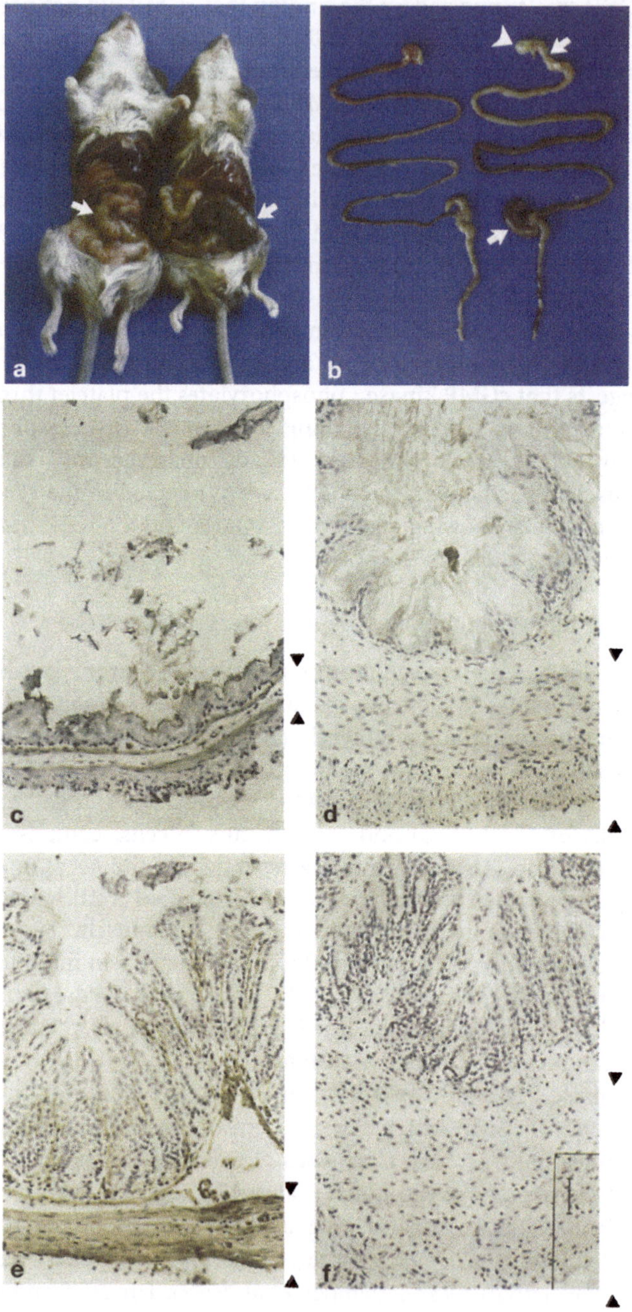
a
b
c
d
e
f

the mutant and wild type mice indicating that the development of the inhibitory enteric nerve system is not affected by the loss of cGMP kinase I. Tissue bath experiments with muscle strips isolated from gastric fundus, pylorus, and ileocaecal junction, clearly demonstrated the block of the NO/cGMP-dependent relaxation of the intestinal smooth muscle (Pfeifer et al. 1998). The deficiency of NO/cGMP-induced relaxation in the cGMP kinase I-deficient smooth muscle leads to defective gastrointestinal relaxation especially of pylorus, ileocecal junction and colon resulting in stasis of food and feces proximal to the constricted segments. X-ray analysis revealed the loss of normal peristalsis which lead to a heavily retarded passage of intestinal content and progressive stasis. The longstanding distension may eventually cause perforation and peritonitis of the small intestine.

The intestinal function is not only affected by a loss of cGMP kinase I but also by that of GMP kinase II. This enzyme is highly expressed in the apical membrane of the enterocytes of the small intestine. Increased intestinal Cl-/fluid secretion is a common feature of substances that stimulate the production of cAMP and cGMP (Field et al. 1989). The heat-stable enterotoxin of *E. coli* (STa) is a potent secretagogue that binds to and activates intestinal guanylyl cyclase C (GC-C) (Schulz et al. 1990; Vaandrager et al. 1993) thereby increasing cellular cGMP. The biological activator of GC-C is the peptide hormon guanylin (Currie et al. 1992), a 15-amino acid peptide with close structural similarity to *E. coli* STa. Recent experiments indicated that both factors increase cGMP and activate cGMP kinase II leading to the phosphorylation of the cystic fibrosis transmembrane conductance regulator (CFTR) and increased Cl^- and water secretion. cGMP kinase II co-localizes with GC-C in the apical membrane of enterocytes (Markert et al. 1995) and activates CFTR Cl- channels in excised patches of CFTR-transfected NIH3T3 fibroblasts and IEC-CF7 cells (French et al. 1995). cGMP kinase II-deficient mucosa isolated from the small intestine responded normally to cAMP analogues, whereas STa-induced electrogenic anion secretion was blocked

Fig. 5A–D. Analysis of the intestinal defects of cGMP kinase I-deficient (cGKI-/-) mice. (**A**) Autopsy of a 7 week old wild type (left) and litter matched cGKI-/- mouse (right). The arrow indicates the position of the caecum. (**B**) Entire gastrointestinal tract dissected from the same mice as in (A). (left) wild type, (right) cGKI-/-. The arrowhead and the upper arrow mark the distended stomach and constricted pylorus, respectively. Together with the poststenotic dilated duodenum, this triad is indicative of pyloric stenosis. The lower arrow indicates the distended caecum of the cGKI-/- mouse. (**C–F**) Immunohistochemical analysis of cGKI expression in fundus (C and D) and pylorus (E and F) isolated from a 4 weeks old wild type (C and E) and littermatched cGKI-/- (D and F) mouse.The arrowheads indicate the width of the muscle layer

(Pfeifer et al. 1996) indicating that loss of cGMP kinase II selectively disrupted the STa/cGMP athway without affecting cAMP signaling.

By use of an *in vivo* model for enterotoxin evaluation ("suckling mice") (Giannella 1981), it was shown that STa induced the accumulation of clear fluid into the intestine and increased the relative gut weight ~2-fold in wild type and heterozygote cGMP kinase II+/- mice. In contrast, no significant accumulation of fluid into the intestine was observed after administration of STa in the cGMP kinase II-/- mice. However, inactivation of the cGMP kinase II gene did not affect the secretory response to cholera toxin which increases cAMP (Pfeifer et al. 1996). Therefore, the experiments conducted in the cGMP kinase II deficient mice clearly show that cGMP kinase II is the physiological mediator of the STa/cGMP effects in the small intestine and that cGMP and cAMP signal independently from each other. The cGMP kinase II-/- mouse is the first animal model that exhibits resistance to *E. coli* STa, and has important clinical implications since STa is a major cause of traveller's diarrhea and has been postulated to be responsible for up to 50% of infant mortality in developing countries (Giannella 1981). The fact, that cGMP kinase II-deficient mice exhibit no gross abnormalities under physiological conditions in their intestine, is not per se contradictory to a protective function of cGMP/cGMP kinase II. Considering the optimum housing conditions in modern animal facilities, a defect in mucosal protection might only be seen after challenging the mice with the adequate enviromental stimulus.

Effects of cGMP kinase on gene transcription and growth

Although a wealth of data on cAMP kinase mediated regulation of gene transcription has been gathered, little is known about the impact of the NO/cGMP signaling cascade on gene expression. Among its diverse actions, NO has been shown to change gene expression and to induce cellular differentiation in hematopoietic cells (Magrinat et al. 1992). Transfection studies using chloramphenicol acetyltransferase (CAT) vectors demonstrated that NO and cGMP analogues activate phorbol ester response element (TRE)-regulated reporter genes (Pilz et al. 1995), a major target of the protein kinase C-signalling pathway. Activation of TRE by cGMP seems to involve the AP-1 (Jun/Fos) transcription factor since DNA binding of AP-1 was increased in cGMP-treated cells (Pilz et al. 1995). In addition, incubation of rat embryo fibroblasts (REF52 cells) with NO-liberating agents and cGMP caused an increase in junB and cfos mRNA expression that was already detectable after 30 min incubation. The time-course of cGMP effects on gene transcription correlated with the phosphorylation of VASP – a major target of cGMP kinase phosphorylation (Halbrügge et al. 1990) – suggesting that cGMP kinase I mediates the effects of NO/cGMP on gene transcription (Pilz et al. 1995). Interestingly, cGMP kinase I contains a putative nuclear localization signal, which is required for the cGMP-induced nuclear translocation of the kinase (Gudi et al. 1997). Subseqent studies using cGMP kinase-deficient cells (BHK cells) and CAT reporter constructs demonstrated a direct role of cGMP kinase I in the regulation of gene expression (Gudi et al. 1996). Transfection of cGMP kinase Iß into BHK cells caused activation of the human *cfos* promotor indicating that the *cfos* gene is directly regulated by cGMP kinase. The product of the protooncogene *cfos* is a central part of the AP-1 transcription factor complex and is involved in regulation of cell growth and differentiation. Although CRE-dependent reporter gene expression was stimulated by cGMP-analogues in BHK cells overexpressing cGMP kinase Iß (Gudi et al. 1996), no significant activation of CRE-regulated reporter genes was found in rat thyroid cells (Pilz et al. 1995) indicating that responsiveness of CRE-containing promotors is cell-type-specific, and that cGMP is able to signal independently from cAMP kinase to the nucleus.

Substantial experimental evidence has been obtained which suggests a pivotal role for cGMP kinase I in vascular remodeling and smooth muscle proliferation. The proliferation of vascular cells is apparently regulated by NO. Garg and Hassid (1989) showed that exogenous NO inhibits the growth of vascular smooth muscle cells. Similarly, NO, ANF, and cGMP inhibit the growth-promoting effects of norepinephrine in cardiac myocytes and fibroblasts (Calderone et al. 1998). The NO dependent inhibition of smooth muscle proliferation is apparently mediated by cGMP kinase I (Yu et al. 1997). The potential target of cGMP kinase I seems to be raf-1 kinase. Phosphorylation of this kinase was reported to inhibit Ras-dependent activation of raf-1 (Yu et al. 1997). Vascular remodelling could be inhibited further by a cGMP kinase I-dependent block of the secretion of the extracellular matrix proteins osteopontin and thrombospondin (Dey et al. 1998).

cGMP kinases and bone growth

cGMP kinase I and II are expressed in the growth zone of bones (Pfeifer et al. 1996). The deletion of cGMP kinase I has no apparent effect on the growth of the skeleton (Pfeifer et al. 1998). In contrast, cGMP kinase II deficient mice are dwarfs with between 16 and 30% shorter limbs. cGMP kinase II plays a direct role in endochondral ossification. The enzyme is expressed in growth plate chondrocytes (Pfeifer et al. 1996) and regulates autonomous bone growth.

The unexpected role of cGMP kinase II in chondrocyte function raises two questions: (i) which signalling cascade ties to cGMP kinase II, and (ii) what are the downstream targets of this kinase? Principally, cGMP may be derived from the soluble or particulate GCs. The inducible NOS has been cloned from human articular chondrocytes (Charles et al. 1993) and the neuronal NOS (nNOS) is expressed in the growth plate indicating that NO and soluble GC may play a role in endochondral ossification. At present, it is not clear whether or not soluble GC is also expressed in growth plate chondrocytes. In contrast, the particulate GCs, GC-A and GC-B, are expressed abundantly in mouse tibial epiphysis and vertebrae (Suda et al. 1998). Cultivation of mouse tibias in the presence of 1μM brain derived natriuretic peptide (BNP) – a ligand for GC-A and GC-B – induced a significant increase in total bone length. Transgenic mice overexpressing BNP exhibit skeletal overgrowth which is restricted to those bones that grow by endochondral ossification. The phenotypes of the BNP-transgenic and the cGMP kinase II deficient mice are opposite. They establish that endochondral ossifaction and normal skeletal growth is regulated by cGMP and cGMP kinase II.

The target(s) of cGMP kinase II in chondrocytes are not known at present. However, the analyses of the BNP-transgenics and cGMP kinase II-knockouts suggest that the cGMP pathway is essential for the proper differentiation of proliferative chondrocytes and/or the function of hypertrophic chondrocytes. In order to increase the cellular volume and to become hypertrophic, chondrocytes must be able to regulate the flux of electrolytes via the cell-membrane. In analogy to the intestine where the CFTR is a major substrate for cGMP kinase II, this or a similar Cl^- channel might also be a target in the growth plate.

cGMP kinases and neuronal functions

Rat cerebellum was the first mammalian organ that was shown to express a cGMP-stimulated protein kinase activity (Hofmann and Sold 1972). Although the exact function of the cerebellum remains controversial, the general hypothesis of cerebellar function is that the cerebellum is a primary site of motor learning. Purkinje cells are the only known output from the cerebellar cortex, and project to the deep cerebellar nuclei (Raymond et al. 1996). cGMP kinase I expression is confined to Purkinje cells indicating that cGMP kinase I might be involved in motor learning processes. The output of Purkinje cells is regulated by two distinct excitatory inputs, the parallel fibers (PFs) and the climbing fibers (CFs). Conjunctive stimulation of PF and CF inputs at low frequencies results in a persistent attenuation of the PF-Purkinje neuron synapse, a process termed long-term depression (LTD) (Ito et al. 1982). Although the participation of NO/cGMP in LTD was already proposed in 1990 (Ito and Karachot 1990), its role remains controversial. All necessary parts of the NO/cGMP/cGMP kinase signalling cascade are available in the Purkinje cells: NO synthase (NOS) is found in the vicinity of

Purkinje cells (Bredt et al. 1990), GC is highly expressed in these cells (Ariano et al. 1982), and they contain high concentrations of cGMP kinase I (Keilbach et al. 1992). Several studies on cerebellar slice preparations suggested that NO is necessary for LTD. Block of endogenous NO production by NOS inhibitors prevents LTD formation and application of NO to the organ bath can substitute for the CF stimulation in rat slice preparations (Shibuki and Okada 1991). In addition, release of caged NO inside the Purkinje cell completely replaces PF activity and synergizes with depolarization to cause LTD (Lev-Ram et al. 1995). Similar experiments with caged cGMP revealed that cGMP can induce LTD (Lev-Ram et al. 1997). The effect of cGMP was prevented by inhibition of cGMP kinase activity and was dependent on a subsequent or coincident increase in $[Ca^{2+}]_i$ (Lev-Ram et al. 1997). In contrast to these experiments, cerebellar LTD is not attenuated in mice deficient either in cGMP kinase I or II. In agreement with this finding are experiments on cultured murine Purkinje neurons showing, that the NO/cGMP pathway is not required for cerebellar LTD (Linden et al. 1995). These finding suggests that cGMP kinase I is not required for the induction of LTD. It is possible, that cGMP signals in the intact animal by a different pathway involving either CNG channels or cAMP.

The hippocampus has been identified as a brain region necessary for spatial lerning. Long-term potentiation (LTP), the sustained increase of synaptic transmission following repetitive activation (Bliss and Collingridge 1993), is an intriguing candidate for a cellular mechanism underlying learning and memory formation. The Schaffer collateral-commissural pathway conveying inputs to pyramidal neurons in the CA1 region of the hippocampus exhibits a form of LTP which critically depends on a primary NMDA receptor-mediated Ca^{2+} influx into the postsynapse and is, at least partly, due to an increase of presynaptic transmitter release (Bolshakov and Siegelbaum 1995). Freely diffusible nitric oxide (NO), generated postsynaptically by Ca^{2+}-calmodulin-dependent NO synthase (NOS), has been identified as the messenger of the underlying retrograde signaling mechanism (Schuman and Madison, 1991; Zhuo et al. 1994; Arancio et al. 1995). cGMP kinase I is expressed in hippocampal neurons (Kleppisch et al. 1998). It was reported that the pharmacological inhibition of cGMP kinase blocks LTP and that LTP was promoted by combining a cGMP kinase activator with a weak tetanic stimulus (Zhuo et al. 1994). Similar observations made at synapses between individual hippocampal pyramidal neurons in culture further supported this concept (Arancio et al. 1995). The functional role of cGMP kinases in LTP was analyzed in mice lacking the gene for cGMP kinase I, cGMP kinase II or both (Kleppisch et al. 1998). LTP was not defective in the mutant mice lineages, although it was reduced in the presence of the NO synthase inhibitior

N^{ω}-nitro-L-arginine (NOArg). In the wild type, the reduction in LTP caused by NOArg was not abolished by the cGMP kinase activator 8p-CPT-cGMP. Also, the inhibitor of sGC, ODQ, had no effect on LTP, although ODQ suppressed NO-induced cGMP production. Thus, genetic evidence in mice suggests that cGMP kinase is not involved in LTP and that NO acts through an alternative, cGMP-independent pathway.

Drosophila kinases and behavior

The search for the gene responsible for the foraging (*for*) polymorphism – a natural behavior polymorphism in *Drosophila* – led to the identification of a mutation in the gene coding for DG2 (*dg2*), and demonstrated for the first time, that a cGMP-dependent protein kinase is involved in regulation of a complex behavior. The *for* gene controls the basical food-search strategy of *Drosophila*, and two contrasting phenotypes have been described: rovers, and sitters, which differ only in the distances they move while feeding (Sokolowski 1980). In natural populations circa twice more rovers exist than sitters (Sokolowski 1980). The mapping of the *for* locus did reveal that *dg2* lies within this chromosomal region. Moreover, Osborne et al. (1997) could even identify mutations within or near *dg2* in sitter flies. Comparison of the cGMP-dependent kinase activity demonstrated that kinase activity correlated with food-search behavior, i.e. low kinase acitvity in sitters, high cGMP-stimulated phosphotransferase activity in rovers. Finally, overexpression of *dg2* in sitter larvae changed their behavior to the rover phenotype.

6
Conclusion

Over the last few years, a wealth of biochemical and functional data has been gathered on mammalian and *Drosophila* cGMP-dependent protein kinases. Mutant and chimeric cGMP kinase proteins made by molecular biology techniques yielded important biochemical informations, such as the function of the amino-terminal domains of cGMP kinase I and II, the identity of the cGMP-binding sites of cGMP kinase I, substrate specificity of the enzymes and structural details of the catalytic center. Genetic approaches proved to be especially useful for the analysis of the biological function of cGMP kinases. Analysis of naturally occuring *Drosophila* variants in food-search behavior, combined with the introduction of transgenes and transposable P elements into the *Drosophila* genome, allowed the identification of the long-sought *foraging* gene as the *dg2* gene. In addition, these studies unequivocally demonstrated that cGMP kinases regulate complex central nervous processes, and that even subtle mutations in cGMP kinases can lead to naturally occurring behavior variants. The production of knockout mouse strains that carry null-mutations of the cGMP kinase I or II is an important step towards unravelling the physiological functions of these two kinases. The analysis of these mice lines showed that in many tissues cAMP and cGMP elicit the same physiological response, but use independent pathways. These mice showed that cGMP kinase II is the biological mediator of the cGMP-induced Cl^-/water secretion into the small intestine and is essential for normal skeletal growth. The phenotypes of the cGMP kinase I-deficient mice clearly identified this enzyme as the physiological mediator of NO/cGMP in many tissues. It was surprising that these mice develop pathological abnormalities at all, given the fact that many effects induced by the cGMP and cAMP pathways are identical. These results demonstrate that, in addition to biochemical and cell biological techniques, only genetic models allow in depth insight into the physiological significance of complicated signaling pathways such as the NO/ANF/cGMP system.

References

Aitken A; Bilham T; Cohen P; Aswad D; Greengard P (1981) A specific substrate from rabbit cerebellum for guanosine-3':5'-monophosphate-dependent protein kinase. III. Amino acid sequences at the two phosphorylation sites. J Biol Chem 256:3501–3506

Aitken A, Hemmings BA, Hofmann F (1984) Identification of the residues on cyclic GMP-dependent protein kinase that are autophosphorylated in the presence of cyclic AMP and cyclic GMP. Biochim Biophys Acta 790:219–225

Alioua A, Tanaka Y, Meera P, Wallner M, Hofmann F, Ruth P, Toro L (1998) Biochemical evidence for direct phosphorylation of the maxiK channel α -subunit (*hslo*) by cGMP-dependent protein kinase (PKG). Biophys J 74:A211

Ally S, Tortora G, Clair T, Grieco D, Merlo G, Katsaros D, Ogreid D, Doskeland SO, Jahnsen T, Cho-Chung-YS (1988) Selective modulation of protein kinase isozymes by the site-selective analog 8-chloroadenosine 3',5'-cyclic monophosphate provides a biological means for control of -human colon cancer cell growth. Proc Nat Am Soc USA 85:6319–6322

Arancio O, Kandel ER, Hawkins RD (1995) Activity-dependent long-term enhancement of transmitter release by presynaptic 3',5'-cyclic GMP in cultured hippocampal neurons. Nature 376:74–80

Ariano MA, Lewicki JA, Brandwein HJ, Murad F (1982) Immunohistochemical localization of guanylate cyclase within neurons of rat brain. Proc Natl Acad Sci USA 79:1316–1320

Atkinson RA, Saudek V, Huggins JP, Pelton JT (1991) 1H NMR and circular dichroism studies of the N-terminal domain of cyclic GMP dependent protein kinase: a leucine/isoleucine zipper. Biochemistry 30:9387–9395

Biel M, Zong X, Ludwig A, Sautter A, Hofmann F (1998) Structure and function of cyclic nucleotide-gated channels. Rev Physiol Biochem Pharmacol Chapter 5

Bliss TVP, Collinridge GL (1993) A synaptic model of memory: long-term potentiation in the hippocampus. Nature 361:31–39

Bolshakov VV, Siegelbaum SA (1995) Regulation of hippocampal transmitter release during development and long-term potentiation. Science 269:1730–1734

Bossemeyer D, Engh RA, Kinzel V, Ponstingl H, Huber R (1993) Phosphotransferase and substrate binding mechanism of the cAMP-dependent protein kinase catalytic subunit from porcine heart as deduced from the 2.0 A structure of the complex with Mn^{2+} adenylyl imidodiphosphate and inhibitor peptide PKI(5–24). EMBO J 12:849–859

Bredt DS, Hwang PM, Snyder SH (1990) Localization of nitric oxide synthase indicating a neural role for nitric oxide. Nature 347:768–770

Burns AJ, Lomax AEJ, Torihashi S, Sanders KM, Ward SM (1996) Interstitial cells of Cajal mediate inhibitory neurotransmission. Proc Natl Acad Sci USA 93:12008–12013

Butt E, Abel K, Krieger M, Palm D, Hoppe V, Hoppe J, Walter U (1994) cAMP- and cGMP-dependent protein kinase phosphorylation sites of the focal adhesion vasodilator-stimulated phosphoprotein (VASP) in vitro and in intact human platelets. J Biol Chem 269:14509–14517

Butt E, Pöhler D, Genieser HG, Huggins JP, Bucher B (1995) Inhibition of cyclic GMP-dependent protein kinase-mediated effects by (Rp)-8-bromo-PET-cyclic GMPS. Br J Pharmacol 116:3110–3116

Calderone A, Thaik CM, Takahashi N, Chang DLF, Colucci WS (1998) Nitric oxide, atrial natriuretic peptide, and cyclic GMP inhibit the growth-promoting effects of norepinephrine in cardiac myocytes and fibroblasts. J Clin Invest 101:812–818

Casey PJ (1995) Protein lipidation in cell signaling. Science 268:221–225

Chao DS, Silvagno F, Xia H, Cornwell TL, Lincoln TM, Bredt DS (1997) Nitric oxide synthase and cyclic GMP-dependent protein kinase concentrated at the neuromuscular endplate. Neuroscience 76:665–672

Charles IG, Palmer RM, Hickery MS, Bayliss MT, Chubb AP, Hall VS, Moss DW, Moncada S (1993) Cloning, characterization, and expression of a cDNA encoding an inducible nitric oxide synthase from the human chondrocyte. Proc Natl Acad Sci USA 90:11419–11423

Chestukhin A, Litovchick L, Schourov D, Cox S, Taylor SS, Shaltiel S (1996) Functional malleability of the carboxyl-terminal tail in protein kinase A. J Biol Chem 271:10175–10182

Colbran JL, Francis SH, Leach AB, Thomas MK, Jiang H, McAllister LM, Corbin JD (1992) A phenylalanine in peptide substrates provides for selectivity between cGMP- and cAMP-dependent protein kinases. J Biol Chem 267:9589–9594

Corbin JD, Ogreid D, Miller JP, Suva RH, Jastorff B, Doskeland SO (1986) Studies of cGMP analog specificity and function of the two intrasubunit binding sites of cGMP-dependent protein kinase. J Biol Chem 261:1208–121

Cornwell TL, Lincoln TM (1989) Regulation of intracellular Ca^{2+} levels in cultured vascular smooth muscle cells. J Biol Chem 264:1146–1155

Currie MG, Fok KF, Kato J, Moore RJ, Hamra FK, Duffin KL, Smith CE (1992) Guanylin: an endogenous activator of intestinal guanylate cyclase. Proc Natl Acad Sci USA 89:947–951

De Jonge HR (1981) Cyclic GMP-dependent protein kinase in intestinal brush borders. Adv Cyclic Nucleotide Res 14:315–333

Desai KM, Zembowicz A, Sessa WC, Vane JR (1991) Nitroxergic nerves mediate vagally induced relaxation in the isolated stomach of the guinea pig. Proc Natl Acad Sci USA 88:11490–11494

Dey NB, Boerth NJ, Murphy-Ullrich JE, Chang PL, Prince CW, Lincoln TM (1998) Cyclic GMP-dependent protein kinase inhibits osteopontin and thrombospondin production in rat aortic smooth muscle cells. Circ Res 82:139–146

Diwan AH, Thompson WJ, Lee AK, Strada SJ (1994) Cyclic GMP-dependent protein kinase activity in rat pulmonary microvascular endothelial cells. Biochem Biophys Res Commun 202:728–735

Dostmann WRG (1995) (Rp)-cAMPS inhibits the cAMP-dependent protein kinase by blocking the cAMP-induced conformational transition. FEBS Lett 375:231–234

Dostmann WRG, Taylor SS (1991) Identifying the molecular switches that determine whether (Rp)-cAMPS functions as an antagonist or an agonist in the activation of cAMP-dependent protein kinase. Biochemistry 30:8710–8716.

Dostmann WRG, Taylor SS, Genieser HG, Jastorff B, Doskeland SO, Ogreid D (1989) Probing the cyclic nucleotide binding sites of cAMP-dependent protein kinases I and II with analogs of adenosine 3',5'-cyclic phosphorothioates. J Biol Chem 265:10484–10491

Draijer R, Vaandrager AB, Nolte C, De Jonge HR, Walter U, van Hinsbergh VW (1995) Expression of cGMP-dependent protein kinase I and phosphorylation of its substrate, vasodilator-stimulated phosphoprotein, in human endothelial cells of different origin. Circ Res 77:897–905

Eigenthaler M, Ullrich H, Geiger J, Horstrup K, Honig-Liedl P, Wiebecke D, Wlater U (1993) Defective nitrovasodilator-stimulated protein phosphorylation and calcium regulation in cGMP-dependent protein kinase-deficient human platelts of chronic myelocytic leukemia. J Biol Chem 268:13526–13531

El-Daher SS, Eigenthaler M, Walter U, Furuichi T, Miyawaki A, Mikoshiba K, Kakkar V, Authi K (1996) Distribution and activation of cAMP- and cGMP-depedent protein kinases in highly purified human platelet plasma and intracellular membranes. Thromb. Haemostasis 76:1063–1071

El-Husseini AE, Bladen C, Vincent SR (1995) Molecular characterization of a type II cyclic GMP-dependent protein kinase expressed in the rat brain. J Neurochem 64:2814–2817

Eliasson MJL, Blackshaw S, Schell MJ, Snyder SH (1997) Neuronal nitric oxide synthase alternatively spliced forms: prominent functional localizations in the brain. Proc Natl Acad Sci USA 94:3396–3401

Engh RA, Girod A, Kinzel V, Huber R, Bossemeyer D (1996) Crystal structures of catalytic subunit of cAMP-dependent protein kinase in complex with isoquinolinesulfonyl protein kinase inhibitors H7, H8, and H89. Structural implications for selectivity. J Biol Chem 271:26157–26164

Feil R, Kellermann J, Hofmann F (1995) Functional cGMP-dependent protein kinase is phosphorylated in its catalytic domain at threonine-516. Biochemistry 34:13152–13158

Felbel J, Trockur B, Ecker T, Landgraf W, Hofmann F (1988) Regulation of cytosolic calcium by cAMP and cGMP in freshly isolated smooth muscle cells from bovine trachea. J Biol Chem 263:16764–16771

Field M, Rao MC, Chang EB (1989) Intestinal electrolyte transport and diarrheal disease (1) N Engl J Med 321:800–806

Foster JL, Higgins GC, Jackson FR (1996) Biochemical properties and cellular localization of the Drosophila DG1 cGMP-dependent protein kinase. J Biol Chem 271:23322–23328

Francis SH, Noblett BD, Todd BW, Wells JN, Corbin JD (1988) Relaxation of vascular and tracheal smooth muscle by cyclic nucleotide analogues that preferentially activate purified cGMP-dependent protein kinase. Mol Pharmacol 34:506–517

French PJ, Bijman J, Edixhoven M, Vaandrager AB, Scholte BJ, Lohmann SM, Nairn AC, de Jonge HR (1995) Isotype-specific activation of cystic fibrosis transmembrane conductance regulator-chloride channels by cGMP-dependent protein kinase II. J Biol Chem 270:26626–26631

Fujii M, Ogata T, Takahashi E, Yamada K, Nakabayashi K, Oishi M, Ayusawa D (1995) Expression of the human cGMP-dependent protein kinase II gene is lost upon introduction of SV40 T antigen or immortalization in human cells. FEBS Lett 375:263–267

Gambaryan S, Häusler C, Markert T, Pohler D, Jarchau T, Walter U, Haase W, Kurtz A, Lohmann SM (1996) Expression of type II cGMP-dependent protein kinase in rat kidney is regulated by dehydration and correlated with renin gene expression. J Clin Invest 98:662–670

Gamm DM, Francis SH, Angelotti TP, Corbin JD, Uhler MD (1995) The type II isoform of cGMP-dependent protein kinase is dimeric and possesses regulatory and catalytic properties distinct from the type I isoforms. J Biol Chem 270:27380–27388

Gardes J, Poux JM, Gonzalez MF, Alhenc-Gelas F, Menard J (1992) Decreased renin release and constant kallikrein secretion after injection of L-NAME in isolated perfused rat kidney. Life Sci 50:987–993

Garg UC, Hassid A, (1989) Nitric oxide-generating vasodilators and 8-bromo-cyclic guanosine monophosphate inhibit mitogenesis and proliferation of cultured rat vascular smooth muscle cells. J Clin Invest 83:1774–1777

Geiger J, Nolte C, Butt E, Sage SO, Walter U (1992) Role of cGMP and cGMP-dependent protein kinase in nitrovasodilator inhibition of agonist-evoked calcium elevation in human platelets. Pròc Natl Acad Sci USA 89:1031–1035

Giannella RA (1981) Pathogenesis of acute bacterial diarrheal disorders. Ann Rev Med 32:341–357

Gjertsen BT, Mellgren G, Otten A, Maronde E, Genieser HG, Jastorff B, Vintermyr OK, McKnight GS, Doskeland SO (1995) Novel (Rp)-cAMPS analogs as tools for inhibition of cAMP-kinase in cell culture. Basal cAMP-kinase activity modulates interleukin-1 beta action. J Biol Chem 270:20599–20607

Glass DB (1983) Differential responses of cyclic GMP-dependent and cyclic AMP-dependent protein kinases to synthetic peptide inhibitors. Biochem J 213:159–164

Glass DB, Cheng HC, Kemp BE, Walsh DA (1986) Differential and common recognition of the catalytic sites of the cGMP-dependent and cAMP-dependent protein kinases by inhibitory peptides derived from the heat-stable inhibitor protein. J Biol Chem 261:12166–12171

Glass DB, Feller MJ, Levin LR, Walsh DA (1992) Structural basis for the low affinities of yeast cAMP-dependent and mammalian cGMP-dependent protein kinases for protein kinase inhibitor peptides. Biochemistry 31:1728–1734

Glass DB, Krebs EG (1982) Phosphorylation by guanosine 3':5'-monophosphate-dependent protein kinase of synthetic peptide analogs of a site phosphorylated in histone H2B. J Biol Chem 257:1196–200

Grant PG, Mannarino AF, Colman RW (1990) Purification and characterization of a cyclic GMP-stimulated cyclic nucleotide phosphodiesterase from the cytosol of human platelets. Thromb Res 59:105–119
Greenberg SG, He XR, Schnermann JB, Briggs JP (1995) Effect of nitric oxide on renin secretion. I. Studies in isolated juxtaglomerular granular cells. Am J Physiol 268:F948–F952
Greenstein-Baynash A, Hosoda K, Giaid A, Richardson JA, Emoto N, Hammer RE, Yanagisawa M (1994) Interaction of enothelin-3 with endothelin-b receptor is essential for development of epidermal melanocytes and enteric neurons. Cell 79:1277–1285
Grunberg B; Negrescu E; Siess.W (1995) Synergistic phosphorylation of platelet rap1B by SIN-1 and iloprost. Eur J Pharmacol 288:329–333
Gudi T, Huvar I, Meinecke M, Lohmann SM, Boss GR, Pilz RB (1996) Regulation of gene expression by cGMP-dependent protein kinase. Transactivation of the c-fos promoter. J Biol Chem 271:4597–4600
Gudi T, Lohmann SM, Pilz RB (1997) Regulation of gene expression by cyclic GMP-dependent protein kinase requires nuclear translocation of the kinase: identification of a nuclear localization signal. Mol Cell Biol 17:5244–5254
Halbrügge M, Friedrich C, Eigenthaler M, Schanzenbacher P, Walter U (1990) Stoichiometric and reversible phophorylation of a 46-kD protein in human platelets in response to cGMP- and cAMP-elevating vasodilators. J Biol Chem 265:3088–3093
Hanks SK, Hunter T (1995) The eukaryotic protein kinase superfamily: kinase (catalytic) domain structure and classification. FASEB J 9:576–596
Hashimoto E; Takeda M; Nishizuka Y; Hamana K; Iwai K (1976) Studies on the sites in histones phosphorylated by adenosine 3':5'-monophosphate-dependent and guanosine 3':5'-monophosphate-dependent protein kinases. J Biol Chem 251:6287–6293
He XR, Greenberg SG, Briggs JP, Schnermann JB (1995) Effect of nitric oxide on renin secretion. II. Studies in the perfused juxtaglomerular apparatus. Am J Physiol 268:F953–F959
Heil WG, Landgraf W, Hofmann F (1987) A catalytically active fragment of cGMP-dependent protein kinase. Occupation of its cGMP-binding sites does not affect its phosphotransferase activity. Eur J Biochem 168:117–121
Henrich WL, McAllister EA, Smith PB, Campbell WB (1988) Guanosine 3',5'-cyclic monophosphate as a mediator of inhibition of renin release. Am J Physiol 255:F474–F478
Herberg FW, Taylor SS, Dostmann WRG (1996) Active site mutations define the pathway for the cooperative activation of cAMP-dependent protein kinase. Biochemistry 35:2934–2942
Hidaka H, Kobayashi R (1992) Pharmacology of protein kinase inhibitors. Annu Rev Pharmacol Toxicol 32:377–397
Hofmann F, Dostmann W, Keilbach A, Landgraf W, Ruth P (1992) Structure and physiological role of cGMP-dependent protein kinase. Biochim Biophys Acta 1135:51–60
Hofmann F, Gensheimer HP, Göbel, C (1985) cGMP-dependent protein kinase: Autophosphorylation changes the characteristics of binding site 1. Eur J Biochem 147:361–365
Hofmann F, Ludwig A, Pfeifer A (1994) Cyclic GMP and the cotrol of airways smooth muscle tone. Airways Smooth Muscle: Biochemical Control of Contraction and

Relaxation, D. Raeburn and M.A. Giembycz (eds); Birkhäuser Verlag Basel/Switzerland, 253–265

Hofmann F, Sold G (1972) A protein kinase activity from rat cerebellum stimulated by guanosine-3':5'-monophosphate. Biochem Biophys Res Commun 49:1100–1107

Hosoda K, Hammer RE, Richardson JA, Greenstein-Baynash A, Cheung JC, Giaid A, Yanagisawa M (1994) Targeted and natural (Piebald-lethal) mutations of endothelin-b receptor gene produce megacolon associated with spotted coat color in mice. Cell 79:1267–1276

Huang PL, Dawson TM, Bredt DS, Snyder S, Fishman M (1993) Targeted disruption of the neuronal nitric oxide synthase gene. Cell 75:1273–1286

Huang PL, Huang Z, Mashimo H, Bloch KD, Moskowitz MA, Bevan JA, Fishman M (1995) Hypertension in mice lacking the gene for endothelial nitric oxide synthase. Nature 377:239–242

Huizinga JD, Thuneberg L, Klüppel M, Malysz J, Mikkelsen HB, Bernstein A (1995) W/kit gene required for interstitial cells of Cajal and for intestinal pacemaker activity. Nature 373:347–349

Ito M, Karachot L (1990) Messengers mediating long-term desensitization in cerebellar Purkinje cells. Neuroreport 1:129–132

Ito M, Sakurai M, Tongroach P (1982) Climbing fibre induced depression of both mossy fibre responsiveness and glutamate sensitivity of cerebellar Purkinje cells. J Physiol Lond 324:113–134

Jarchau T, Häusler C, Markert T, Pöhler D, Vandekerckhove J, De Jonge HR, Lohmann S, Walter U (1994) Cloning, expression, and in situ localization of rat intestinal cGMP-dependent protein kinase II. Proc Natl Acad Sci USA 91:9426–9430

Jiang H, Colbran JL, Francis SH, Corbin JD (1992) Direct evidence for cross-activation of cGMP-dependent protein kinase by cAMP in pig coronary arteries. J Biol Chem 267:1015–1019

Joyce NC, DeCamilli P, Lohmann SM, Walter U (1986) cGMP-dependent protein kinase is present in high concentrations in contractile cells of the kidney vasculature. J Cyclic Nucleotide Protein Phosphor Res 11:191–198

Kalderon D, Rubin GM (1989) cGMP-dependent protein kinase genes in Drosophila. J Biol Chem 264:10738–10748

Kase H, Iwahashi K, Nakanishi S, Matsuda Y, Yamada K, Takahashi M, Murakata C, Sato A, Kaneko M (1987) K-252 compounds, novel and potent inhibitors of protein kinase C and cyclic nucleotide-dependent protein kinases. Biochem Biophys Res Commun 142:436–440

Keilbach A, Ruth P, Hofmann F (1992) Detection of cGMP-dependent protein kinase isozymes by specific antibodies. Eur J Biochem 208:467–473

Kemp BE, Pearson RB (1991) Intrasteric regulation of protein kinases and phosphatases. Biochim Biophys Acta 1094:67–76

Kleppisch T, Pfeifer A, Klatt P, Ruth P, Montkowski A, Fässler R, Hofmann F (1998) Long-term potentiation in the hippocampal CA1 region of mice lacking the cGMP-dependent protein kinase is normal and susceptible to inhibition of NO synthase J Neurosci (in press)

Knighton DR, Zheng JH, Ten Eyck LF, Ashford VA, Xuong NH, Taylor SS, Sowadski JM (1991) Crystal structure of the catalytic subunit of cyclic adenosine monophosphate-dependent protein kinase. Science 253:407–414

Komalavilas P, Lincoln TM (1994) Phosphorylation of the inositol 1,4,5-trisphosphate receptor by cyclic GMP-dependent protein kinase. J Biol Chem 269:8701–8707

Komalavilas P; Lincoln TM (1996) Phosphorylation of the inositol 1,4,5-trisphosphate receptor. Cyclic GMP-dependent protein kinase mediates cAMP and cGMP dependent phosphorylation in the intact rat aorta. J Biol Chem 271:21933–21938

Kume H, Hall IP, Washabau RJ, Takagi K, Kotlikoff MI (1994) Beta-adrenergic agonists regulate KCa channels in airway smooth muscle by cAMP-dependent and -independent mechanisms.J Clin Invest 93:371–379

Kume H, Tokuma H, Tomita T (1989) Regulation of Ca^{2+}-dependent K^{+}-channel activity in tracheal myocytes by phophorylation. Nature 341:152–154

Kuo JF, Greengard P (1970) Isolation and partial purification of a protein kinase activated by guanosine 3',5'-monophosphate. J Biol Chem 245:2493–2498

Kurtz A, Della-Bruna R, Pfeilschifter J, Taugner R, Bauer C (1986) Atrial natriuretic peptide inhibits renin release from juxtaglomerular cells by a cGMP-mediated process. Proc Natl Acad Sci USA 83:4769–4773

Landgraf W, Hofmann F (1989) The amino terminus regulates binding to and activation of cGMP-dependent protein kinase. Eur J Biochem 181:643–650

Landgraf W, Hofmann F, Pelton JT, Huggins JP (1990) Effects of cyclic GMP on the secondary structure of cyclic GMP dependent protein kinase and analysis of the enzyme's amino-terminal domain by far-ultraviolet circular dichroism. Biochemistry 29:9921–9928

Landgraf W, Hullin R, Göbel C, Hofmann F (1986) Phosphorylation of cGMP-dependent protein kinase increases the affinity for cyclic AMP. Eur J Biochem 154:113–117

Lang D, Lewis MJ (1989) Endothelium-derived relaxing factor inhibits the formation of inositol trisphosphate by rabbit aorta. J Physiol 441:45–52

Lev-Ram V, Jiang T, Wood J, Lawrence DS, Tsien RY (1997) Synergies and coincidence requirements between NO, cGMP, and Ca^{2+} in the induction of cerebellat long-term depression. Neuron 18:1025–1038

Lev-Ram V, Makings LR, Keitz PF, Kao JPY, Tsien RY (1995) Long-term depression in cerebellar Purkinje neurons results from coincidence of nitric oxide and depolarization. Neuron 15:407–415

Lincoln TM, Cornwell LT, Taylor AE (1990) cGMP-dependent protein kinase mediates the reduction of Ca^{2+} by cAMP in vascular smooth muscle cells. Am J Physiol 258:C399–C407

Lincoln TM, Komalavilas P, Mac-Millan-Crow LA, Cornwell TL (1995) cGMP signaling through cAMP- and cGMP-dependent protein kinases. Adv Pharmacol 34:305–322

Lincoln TM, Pryzwansky KB, Cornwell TL, Wyatt TA, MacMillan LA (1993). cyclic GMP-dependent protein kinase in smooth muscle and neutrophils. Adv Second Messenger Phosphoprotein Res 28:121–132

Linden DJ, Dawson TM, Dawson VL (1995) An evaluation of the nitric oxide/cGMP/cGM-dependent protein kinase cascade in the induction of cerebellar long-term depression in culture. J Neurosci 15:5098–5105

Lohmann SM, Walter U, Miller PE, Greengard P, De-Camilli P (1981) Immunohistochemical localization of cyclic GMP-dependent protein kinase in mammalian brain. Proc Natl Acad Sci USA 78:653–657

MacMillan-Crow LA, Lincoln TM (1994) High-affinity binding and localization of the cyclic GMP-dependent protein kinase with the intermediate filament protein vimentin. Biochemistry 33:8035–8043

MacMillan-Crow LA, Murphy Ullrich JE, Lincoln TM (1994) Identification and possible localization of cGMP-dependent protein kinase in bovine aortic endothelial cells. Biochem Biophys Res Commun 201:531–537

Macphee CH, Reifsnyder DH, Moore TA, Lerea KM, Beavo JA (1988) Phosphorylation results in activation of a cAMP phosphodiesterase in human platelets. J Biol Chem 263:10353–10358

Maeda H, Yamagata A, Nishikawa S, Yoshinaga K, Kobayashi S, Nishi K, Nishikawa S-I (1992) Requirement of c-kit for development of intestinal pacemaker system. Development 116:369–375

Magrinat G, Mason SN, Shami PJ, Weinberg JB (1992) Nitric oxide modulation of human leukemia cell differentiation and gene expression. Blood 80:1880–1884

Markert T, Vaandrager AB, Gambaryan S, Pohler D, Hausler C, Walter U, De Jonge HR, Jarchau T, Lohmann SM (1995) Endogenous expression of type II cGMP-dependent protein kinase mRNA and protein in rat intestine. Implications for cystic fibrosis transmembrane conductance regulator. J Clin Invest 96:822–830

McDonald BJ, Moss SJ (1994) Differential phosphorylation of intracellular domains of γ-aminobutyric acid type A receptor subunits by calcium/calmodulin type 2-dependent protein kinase and cGMP-dependent protein kinase. J Biol Chem 269:18111–18117.

McDonald BJ, Moss SJ (1997) Conserved phosphorylation of the intracellular domains of $GABA_A$ receptor β_2 and β_3 subunits by cAMP-dependent protein kinase, cGMP-dependent protein kinase C and Ca^{2+}/calmodulin type II-dependent protein kinase. Neuropharmacology 36:1377–1385

Miglietta LA, Nelson DL (1988) A novel cGMP-dependent protein kinase from Paramecium. J Biol Chem 263:16096–16105

Mitchell RD, Glass DB, Wong CW, Angelos KL, Walsh DA (1995) Heat-stable inhibitor protein derived peptide substrate analogs: phosphorylation by cAMP-dependent and cGMP-dependent protein kinases. Biochemistry 34:528–534

Miura Y; Kaibuchi K; Itoh T; Corbin JD; Francis SH; Takai Y (1992) Phosphorylation of smg p21B/rap1B p21 by cyclic GMP-dependent protein kinase. FEBS Lett 297:171–174

Murofushi H (1974) Protein kinases in Tetrahymena cilia. II. Partial purification and characterization of adenosine 3',5'-monophosphate-dependent and guanosine 3',5'-monophosphate-dependent protein kinases. Biochim Biophys Acta 370:130–139

Ogreid D, Ekanger R, Suva RH, Miller JP, Doskeland SO (1989) Comparison of the two classes of binding sites (A and B) of type I and type II cyclic-AMP-dependent protein kinases by using cyclic nucleotide analogs. Eur J Biochem 181:19–31

Orstavik S, Solberg R, Tasken K, Nordahl M, Altherr MR, Hansson V, Jahnsen T, Sandberg M (1996) Molecular cloning, cDNA structure, and chromosomal localization of the human type II cGMP-dependent protein kinase. Biochem Biophys Res Commun 220:759–765

Orstavik S; Sandberg M; Berube D; Natarajan V; Simard J; Walter U; Gagne R ; Hansson V; Jahnsen T (1992) Localization of the human gene for the type I cyclic GMP-dependent protein kinase to chromosome 10. Cytogenet-Cell-Genet., 59, 270–273

Osborne KA, Robichon A, Burgess E, Butland S, Shaw RA, Coulthard A, Pereira HS, Greenspan RJ, Sokolowski MB (1997) Natural behavior polymorphism due to a cGMP-dependent protein kinase of *Drosophila*. Science 277:834–836

Pfeifer A, Aszódi A, Seidler U, Ruth P, Hofmann F, Fässler R (1996). Intestinal secretory defects and dwarfism in mice lacking cGMP-dependent protein kinase II. Science 274:2082–2086

Pfeifer A, Klatt P, Massberg S, Ny L, Sausbier M, Hirneiß C, Wang G-X, Korth M, Aszódi A, Andersson K-E, Krombach F, Mayerhofer A, Ruth P, Fässler R, Hofmann F (1998) Defective smooth muscle regulation in cGMP kinase I-deficient mice. EMBO J 17:3045–3051

Pfitzer G, Hofmann F, DiSalvo J, Rüegg JC (1984) cGMP and cAMP inhibit tension development in skinned coronary arteries. Plügers Arch 401:277–280

Pilz RB, Suhasini M, Idriss S, Meinkoth JL, Boss GR (1995) Nitric oxide and cGMP analogues activate transcription from AP-1-responsive promoters in mammalian cells. FASEB J 9:552–558

Pöhler D, Butt E, Meissner J, Muller S, Lohse M, Walter U, Lohmann SM, Jarchau T (1995) Expression, purification, and characterization of the cGMP-dependent protein kinases I beta and II using the baculovirus system. FEBS Lett 374:419–425

Pryzwansky KB, Kidao S, Wyatt TA, Reed W, Lincoln TM (1995) Localization of cyclic GMP-dependent protein kinase in human mononuclear phagocytes. J Leukoc Biol 57:670–678

Qian Y, Chao DS, Santillano DR, Cornwell TL, Nairn AC, Greengard P, Lincoln TM, Bredt DS (1996) cGMP-dependent protein kinase in dorsal root ganglion: relationship with nitric oxide synthase and nociceptive neurons. J Neurosci 16:3130–3188

Rannels SR, Corbin JD (1981) Two different intrachain cAMP binding sites of cAMP-dependent protein kinases. J Biol Chem 255:7085–7088

Rapoport RM (1989) Cylic guanosine monophosphate inhibition of contraction may be mediated through inhibition of phosphatidyl inositol hydrolysis in rat aorta. Circ Res 58:407–410

Rapoport RM, Draznin MB, Murad F (1982) Sodium nitroprusside-induced protein phosphorylation in intact rat aorta is mimicked by 8-bromo cyclic GMP. Proc Natl Acad Sci USA 79:6470–6474

Raymond JL, Lisberger SG, Mauk MD (1996) The cerebellum: a neuronal learning machine. Science 272:1126–1131

Reed RB, Sandberg M, Jahnsen T, Lohmann S, Francis S, Corbin J (1996) Fast and slow cyclic nucleotide-dissociation sites in cAMP-dependent protein kinase are transposed in type Iß cGMP-dependent protein kinase. J Biol Chem 271:17570–17575

Reinhard M, Halbrügge M, Scheer U, Wiegand C, Jockusch BM, Walter U (1992) The 46/50 kDa phosphoprotein VASP purified from human platelets is a novel protein associated with actin filaments and focal contacts. Embo J 11:2063–2070

Robertson BE, Schubert R, Hescheler J, Nelson MT (1993) cGMP-dependent protein kinase activates cAMP kinase-activated K+ channels in cerebral artery smooth muscle cells. Am J Physiol 265:C299–C303

Ruth P, Landgraf W, Keilbach A, May B, Egleme C, Hofmann F (1991) The activation of expressed cGMP-dependent protein kinase isozymes I alpha and I beta is determined by the different amino-termini. Eur J Biochem 202:1339–1344

Ruth P, Pfeifer A, Kamm S, Klatt P, Dostmann WRG, Hofmann F (1997). Identification of the amino acid sequences responsible for high affinity activation of cGMP kinase Iα. J Biol Chem 272:10522–10528
Ruth P, Wang G-X, Boekhoff I, May B, Pfeifer A, Penner R, Korth M, Breer H, Hofmann F (1993) Transfected cGMP-dependent protein kinase suppresses calcium transients by inhibition of inositol 1,4,5-triphosphate production. Proc Natl Acad Sci USA 90:2623–2627
Ruth P, Kamm S, Nau U, Pfeifer A, Hofmann F (1996) A cGMP kinase mutant with increased sensitivity to the protein kinase inhibitor peptide PKI(5–24). Biol Chem 377:513–520
Sandberg M, Butt E, Nolte C, Fischer L, Halbrugge M, Beltman J, Jahnsen T, Genieser HG, Jastorff B, Walter U (1991) Characterization of Sp-5,6-dichloro-1-beta-D-ribofuranosylbenzimidazole- 3',5'-monophosphorothioate (Sp-5,6–DCl-cBiMPS) as a potent and specific activator of cyclic-AMP-dependent protein kinase in cell extracts and intact cells. Biochemistry 279:521–527
Sandberg M, Natarajan V, Ronander I, Kalderon D, Walter U, Lohmann S, Jahnsen T (1989) Molecular cloning and predicted full-length amino acid sequence of the type Iß isozyme of cGMP-dependent protein kinase from human placenta. FEBS Lett 255:321–329
Sansom SC, Stockand JD, Hall D, Williams B (1997) Regulation of large calcium-activated potassium channels by protein phosphatase 2A. J Biol Chem 272:9902–9906
Schaap P, van Ments-Cohen M, Soede RDM, Brandt R, Firtel RA, Dostmann W, Genieser HG, Jastorff B, van Haastert PJM (1993) Cell-permeable non-hydrolyzable cAMP derivatives as tools for analysis of signaling pathways controlling gene regulation in Dictyostelium. J Biol Chem 268:6323–6331
Scholz H, Kurtz A (1993) Involvement of endothelium-derived relaxing factor in the pressure control of renin secretion from isolated perfused kidney. J Clin Invest 91:1088–1094
Schulz S, Green CK, Yuen PS, Garbers DL (1990) Guanylyl cyclase is a heat-stable enterotoxin receptor. Cell 63:941–948
Schuman EM, Madison DV (1991) A requirement for the intercellular messenger nitric oxide in long-term potentiation. Science 254:1503–1506
Sekhar KR, Hatchett RJ, Shabb JB, Wolfe L, Francis SH, Wells JN, Jastorff B, Butt E Chakinala MM, Corbin JD (1992) Relaxation of pig coronary arteries by new and potent cGMP analogs that selectively activate type I alpha, compared with type I beta, cGMP-dependent protein kinase. Mol Pharm 42:103–108
Shabb JB, Corbin JD (1992) Cyclic nucleotide-binding domains in proteins having diverse functions. J Biol Chem 267:5723–5726
Shesely EG, Maeda N, Kim HS, Desai KM, Krege JH, Laubach VE, Sherman PA, Sessa WC, Smithies O (1996) Elevated blood pressure in mice lacking endothelial nitric oxide synthase. Proc Natl Acad Sci USA 93:13176–13181
Shesely EG, Maeda N, Kim HS, Desai KM, Krege JH, Laubach VE, Sherman PA, Sessa WC, Smithies O (1996) Elevated blood pressure in mice lacking endothelial nitric oxide synthase. Proc Natl Acad Sci USA 93:13176–13181
Shibuki K, Okada D (1991) Endogenous nitric oxide release required for long-term synaptic depression in the cerebellum. Nature 349:326–328
Shirasawa S, Yunker AMR, Roth KA, Brown GA, Horning S, Korsmeyer SJ (1997) Enx (Hox11L1) deficient mice develop myenteric neuronal hyperplasia and megacolon. Nature Medicine 3:646–650

Sicheri F, Moarefi I, Kuriyan J (1997) Crystal structure of the src family tyrosine kinase hck. Nature 385:602–609
Siess W (1989) Molecular mechanisms of platelet activation. Physiol Rev 69:58–178
Skalhegg BS, Landmark BF, Doskeland SO, Hansson V, Lea T, Jahnsen T (1992) Cyclic AMP-dependent protein kinase type I mediates the inhibitory effects of 3',5'-cyclic adenosine monophosphate on cell replication in human T lymphocytes. J Biol Chem 267:15707–15714
Smith JA, Francis SH, Walsh KA, Kumar S, Corbin JD (1996) Autophosphorylation of type Ibeta cGMP-dependent protein kinase increases basal catalytic activity and enhances allosteric activation by cGMP or cAMP. J Biol Chem 271:20756–20762
Sokolowski MB (1980) Foraging strategies of Drosophila melanogaster: a chromosomal analysis. Behav Genet 10:291–302
Sonnenburg WK, Beavo JA (1994) Cyclic GMP and regulation of cyclic nucleotide hydrolysis. Adv Pharmacol 26:87–114
Steinberg RA, Cauthron RD, Symcox MM, Shuntoh H (1993) Autoactivation of catalytic (Cα) subunit of cyclic AMP-dependent protein kinase by phosphorylation of threonine 197. Mol Cell Biol 13:2332–2341
Steinberg RA, Gorman KB, Øgreid D, Døskeland SO, Weber IT (1991) Mutations that alter charge of type I regulatory subunit and modify activation properties of cyclic AMP-dependent Protein Kinase from Mouse S49 Lymphoma cells. J Biol Chem 266:3547–3553
Su Y, Dostmann WRG, Herberg FW, Durick K, Xuong N-H, Ten Eyck L, Taylor SS Varughese KI (1995) Regulatory subunit of protein kinase A: structure of deletion mutant with cAMP binding domains. Science 269:807–813
Suda M, Ogawa Y, Tanaka K, Tamura N, Yasoda A, Takigawa T, Uehira M, Nishimoto H, Itoh H, Saito Y, Shiota K, Nakao K (1998) Skeletal overgrowth in transgenic mice that overexpress brain natriuretic peptide. Proc Natl Acad Sci 95:2337–2342
Suko J, Maurer-Fogy I, Plank B, Bertel O, Wyskovsky W, Hohenegger M, Hellmann G (1993) Phosphorylation of serine 2843 in ryanodine receptor-calcium release channel of skeletal muscle by cAMP-, cGMP- and CaM-dependent protein kinase. Biochim Biophys Acta 1175:193–206
Takahashi SY, Kageyama T, Ohoka T, Ohnishi E (1974) Guanosine 3',5'-monophosphate-dependent protein kinase from silkworm eggs: Purification and Properties. Insect Biochem 4:429–438
Takio K, Wade RD, Smith SB, Krebs EG, Walsh KA, Titani K (1984) Guanosine cyclic 3',5'-phosphate dependent protein kinase, a chimeric protein homologous with two separate protein families. Biochemistry 23:4207–4218
Taniguchi J, Furukawa K-I, Shigekawa M (1993) Maxi K+ channels are stimulated by cyclic guanosine monophosphate-dependent protein kinase in canine coronary artery smooth muscle cells. Pflügers Arch 423:167–172
Taylor SS, Radzio-Andzelm E (1994) Three protein kinase structures define a common motif. Structure 2:345–55
Tegge W, Frank R, Hofmann F, Dostmann WRG (1995) Determination of cyclic nucleotide-dependent protein kinase substrate specificity by the use of peptide libraries on cellulose paper. Biochemistry 34:10569–10577
Thomas MK, Francis SH, Corbin JD (1990) Substrate- and kinase-directed regulation of phosphorylation of a cGMP-binding phosphodiesterase by cGMP. J Biol Chem 265:14971–14978

Turko IV, Francis SH, Corbin JD (1998) Binding of cGMP to both allosteric sites of cGMP-binding cGMP-specific phosphodiesterase (PDE5) is required for its phosphorylation. Biochem J 329:505–510

Uchida S, Sasaki S, Furukawa T, Hiraoka M, Imai T, Hirata Y, Marumo F (1993) Molecular cloning of a chloride channel that is regulated by dehydration and expressed predominantly in kidney medulla. J Biol Chem 268:3821–3824

Uhler M (1993) Cloning and expression of a novel cyclic GMP-dependent protein kinase from mouse brain. J Biol Chem 268:13586–13591

Vaandrager AB, Edixhoven M, Bot AGM, Kroos MA, Jarchau T, Lohmann SM, Genieser H-G, De Jonge HR (1997) Endogenous Type II cGMP-dependent protein kinase exists as a dimer in membranes and can be functionally distinquished from the type I isoforms. J Biol Chem 272:11816–11823

Vaandrager AB, Ehlert EME, Jarchau T, Lohmann SM, De Jonge HR (1996) N-terminal myristoylation is required for membrane localization of cGMP-dependent protein kinase type II. J Biol Chem 271:7025–7029

Vaandrager AB, Smolenski A, Tilly BC, Houtsmuller AB, Ehlert EME, Bot AGM, Edixhoven M, Boomaars WEM, Lohmann SM, De Jonge HR (1998) Membrane targeting of cGMP-dependent protein kinase is required for cystic fibrosis transmembrane conductance regulator Cl- channel activation. Proc Natl Acad Sci USA 95:1466–1471

Vaandrager AB, van der Wiel E, De Jonge HR (1993) Heat-stable enterotoxin activation of immunopurified guanylyl cyclase C. Modulation by adenine nucleotides. J Biol Chem 268:19598–19603

Van Epps-Fung C, Williams JP, Cornwell TL, Lincoln TM, McDonald JM, Radding W, Blair HC (1994) Regulation of osteoclastic acid secretion by cGMP-dependent protein kinase. Biochem Biophys Res Commun 204:565–571

Vanderwinden J-M, Mailleux P, Schiffmann SN, Vanderhaeghen J-J, De Laet MH (1992) Nitric oxide synthase activity in infantile hypertrophic pyloric stenosis. N Engl J Med 327:511–515

Wagner C, Pfeifer A, Ruth P, Hofmann F, Kurtz A (1998) Inhibitory role of cGMP-kinase II in the control of renin secretion and renin expression J Clin Invest (in press)

Waldmann R, Bauer S, Göbel C, Hofmann F, Jakobs KH, Walter U (1986) Demonstration of cGMP-dependent protein kinase and cGMP-dependent phosphorylation in cell-free extracts of platelets. Eur J Biochem 158:203–208

Wang G-R, Zhu Y, Halushka PV, Lincoln TM, Mendelsohn ME (1998) Mechanism of platelet inhibition by nitric oxide: In vivo phosphorylation of thromboxane receptor by cyclic GMP-dependent protein kinase. Proc Natl Acad Sci USA 95:4888–4893

Wanner R, Wurster B (1990) Cyclic GMP-activated protein kinase from Dictyostelium discoideum. Biochim Biophys Acta 1053:179–184

Weber IT, Shabb JB, Corbin HD (1989) Predicted structures of the cGMP-dependent protein kinase: A key Alanine/threonine difference in evolutionary divergence of cAMP and cGMP binding sites. Biochemistry 28:6122–6127

Weber IT, Steitz TA (1987) Structure of a complex of catabolite gene activator protein and cyclic AMP refined at 2.5 Å resolution. J Mol Biol 198:311–326

Weber IT, Takio K, Titani K, Steitz TA (1982) The cAMP-binding domains of the regulatory subunit of cAMP-dependent protein kinase and the catabolite gene activator protein are homologues. Proc Natl Acad Sci USA 79:7679–7683

Weisskopf MG, Castillo PE, Zalutsky RA, Nicoll RA (1994) Mediation of hippocampal mossy fiber long-term potentiation by cyclic AMP. Science 265:1878–1882
Wen W, Taylor S (1994) High-affinity binding of the heat-stable protein kinase inhibitor to the catalytic subunit of cAMP-dependent protein kinase is selectively abolished by mutation of Arg133. J Biol Chem 269:8423–8430
Wernet W, Flockerzi V, Hofmann F (1989) The cDNA of the two isoforms of bovine cGMP-dependent protein kinase. FEBS Lett 251:191–196
White RE, Lee AB, Shcherbatko AD, Lincoln TM, Schonbrunn A, Armstrong DL (1993) Potassium channel stimulation by natriuretic peptides through cGMP-dependent dephosphorylation. Nature 361:263–266
Wolfe L, Francis SH, Corbin JD (1989) Properties of a cGMP-dependent monomeric protein kinase from bovine aorta. J Biol Chem 264:4157–4162
Wood JS; Yan X; Mendelow M; Corbin D; Francis SH, Lawrence DS (1996) Precision substrate targeting of protein kinases. The cGMP- and cAMP-dependent protein kinases. J Biol Chem 271:174–179
Wyatt TA, Lincoln TM, Pryzwansky KB (1991) Vimentin is transiently co-localized with and phosphorylated by cyclic GMP-dependent protein kinase in formyl-peptide-stimulated neutrophils. J Biol Chem 266:21274–21820
Xu RM, Carmel G, Kuret J, Cheng X (1996) Structural basis for selectivity of the isoquinoline sulfonamide family of protein kinase inhibitors. Proc Natl Acad Sci USA 93:6308–6313
Yan X, Corbin JD, Francis SH, Lawrence,DS (1996) Precision targeting of protein kinases. An affinity label that inactivates the cGMP- but not the cAMP-dependent protein kinase. J Biol Chem 271:1845–1848
Yeaman SJ, Cohen P, Watson DC, Dixon GH (1977) The substrate specificity of adenosine 3':5'-cyclic monophosphate-dependent protein kinase of rabbit skeletal muscle. Biochem J 162:411–421
Yokozaki H, Tortora G, Pepe S, Maronde E, Genieser HG, Jastorff B, Cho-Chung YS (1992) Unhydrolyzable analogues of adenosine 3':5'-monophosphate demonstrating growth inhibition and differentiation in human cancer cells. Cancer Res 52:2504–2508
Yu S-M, Hung L-M, Lin C-C (1997) cGMP-eleating agents suppress proliferation of vascular smooth muscel cells by inhibiting the activation of epidemral growth factor signaling pathway. Circulation 95:1269–1277
Zhao J, Trewhella J, Corbin J, Francis S, Mitchell R, Bushin R, Walsh D (1997) Progressive cyclic nucleotide-induced conformational changes in the cGMP-dependent protein kinase studied by small angle x-ray scattering in solution. J Biol Chem 272:31929–31936
Zheng J, Knighton DR, ten Eyck LF, Karlsson R, Xuong N, Taylor SS, Sowadski JM (1993) Crystal structure of the catalytic subunit of cAMP-dependent protein kinase complexed with MgATP and peptide inhibitor. Biochemistry 32:2154–2161
Zhuo M, Hu Y, Schultz C, Kandel ER, Hawkins RD (1994) Role of guanylyl cyclase and cGMP-dependent protein kinase in long-term potentiation. Nature 368:635–639
Zhou X-B, Ruth P, Schlossmann J, Hofmann F, Korth M (1996) Protein phosphatase 2A is essential for the activation of Ca^{2+}activated K^{+} channels currents by cGMP-dependent protein kinase in tracheal smooth muscle and chinese hamster ovary cells. J Biol Chem 271:19760–19767

Structure and Function of Cyclic Nucleotide-Gated Channels

M. Biel, X. Zong, A. Ludwig, A. Sautter, and F. Hofmann

Institut für Pharmakologie und Toxikologie der Technischen Universität München, Biedersteiner Straße 29, D-80802 München, Germany

Contents

1 Introduction

Cyclic nucleotides are important second messengers that trigger cellular processes such as visual and olfactory transduction (Finn et al. 1996; Zufall et. 1994), synaptic plasticity (Arancio et al. 1995), regulation of smooth muscle tone (Hofmann et al. 1994) and intestinal chloride and water secretion (Vaandrager and De Jonge 1994). Cyclic nucleotides exert their cellular effects by binding to three major classes of cellular receptors, the cyclic nucleotide-dependent protein kinases (Hofmann et al. 1992), the cyclic nucleotide-regulated phosphodiesterases (Beavo 1995) and the cyclic nucleotide-gated (CNG) cation channels. The first identified members of the CNG channel family were the cGMP activated cation channels of rod (Fesenko et al. 1985; Yau and Nakatani 1985) and cone (Cobbs et al. 1985; Haynes et al. 1985) photoreceptors. A CNG channel with a roughly equal sensitivity to both cGMP and cAMP was detected in the membrane of olfactory neurons (Nakamura and Gold 1987). For some years it was thought that CNG channel expression is limited to sensory cells and that the basic function of CNG channels consists in generating receptor currents by allowing influx of Na^+ upon activation by cAMP or cGMP. However, the work of several groups has provided evidence that this physiological function might be restricted to vertebrate photoreceptors. It has become clear, that CNG channels are present in many cell types and that they represent receptor-operated calcium channels which allow neurotransmitter-controlled calcium entry.

2 CNG Channel Genes

CNG channels are key elements in visual (Baylor 1996) and olfactory (Breer et al. 1994) transduction. Both signaling pathways differ from each other in the intrinsic second messenger and the type of CNG channel involved. Visual transduction of rod and cone photoreceptors is controlled by the concentration of cGMP in the outer segment of both photoreceptor types. In mammalian photoreceptors, a high concentration of cGMP is present in the dark maintaining the CNG channel in the open state and leading to a membrane depolarization. Light induces a hyperpolarization of the photoreceptor membrane by triggering the hydrolysis of cGMP and, thereby, the closure of the CNG channel. Interestingly, photoreceptors of the lizard parietal-eye respond to light by a depolarization instead of a hyperpolarization (Finn et al. 1997). The underlying transduction cascade involves the light-induced synthesis of cGMP followed by the activation of a cGMP-activated cation channel that may be related to rod and cone photoreceptor CNG channels.

Olfactory signal transduction is coupled to the odor-dependent synthesis of cAMP which is catalyzed by the adenylyl cyclase of olfactory cilia (Zufall et al. 1994; Breer et al. 1994). The increase in cAMP concentration opens a cAMP-gated channel leading to a depolarization of the olfactory neuron and the generation of an action potential. It has been suggested that inositol-1,4,5-trisphosphate (IP_3) may also be involved in the transduction of certain odors (Breer et al. 1994; Hatt and Ache 1994). However, at least in mammals the IP_3-mediated pathway of olfaction may be less important or even absent since the deletion of the cAMP-gated channel gene in mice resulted in a complete loss of excitatory olfactory signal transduction (Brunet et al. 1996).

The CNG channel from rod photoreceptor has been purified (Cook et al. 1987) and cloned (Kaupp et al. 1989) from bovine rod outer segments. cDNAs of other CNG channels have been isolated by homology screening using the cDNA of the rod channel or by PCR-based cloning (for a recent review see Biel et al. 1996a). Native CNG channels are oligomeric complexes consisting of distinct, yet homologous subunits (Fig. 1). Presently, the CNG

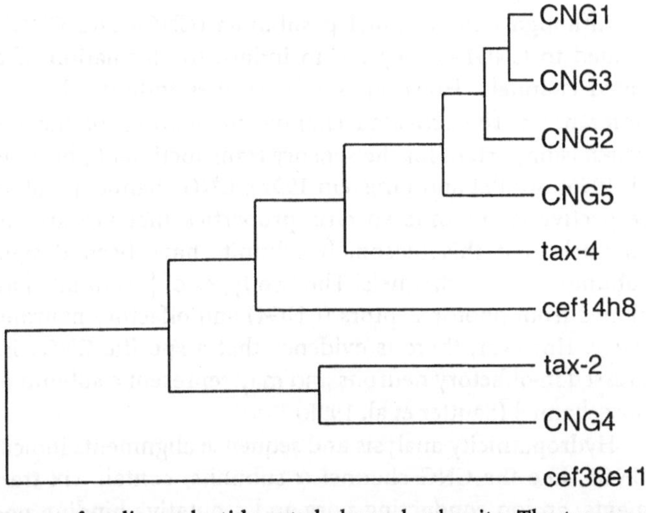

Fig. 1. Phylogenetic tree of cyclic nucleotide-gated channel subunits. The tree was calculated on the basis of sequence alignments with the transmembrane segments and the cyclic nucleotide-binding pocket of the respective subunits. The sequences of mammalian subunits are derived from cattle [CNG1 (Kaupp et al. 1989); CNG3 (Biel et al. 1994); CNG4 (Biel et al. 1996b)] or rat [CNG2 (Dhallan et al. 1990); CNG5 (Bradley et al. 1994], respectively. Tax-4 (Komatsu et al. 1996) and tax-2 (Coburn et al. 1996) represent the α and β subunits of a CNG channel from *C. elegans*. Cef14h8 and cef38e11 represent cosmids containing genomic sequences of *C. elegans* that encode putative CNG channels

channel family comprises five different members in mammals (CNG1–5). The primary sequences of CNG1–5 are homologous to each other indicating that these subunits have evolved from a common ancestral channel. CNG channel subunits can be classified into two different groups on the basis of their ability (α subunits) or inability (β subunits) to form functional homomeric channels in heterologous expression systems. The α subunits of the rod photoreceptor (CNG1), cone photoreceptor (CNG3) and olfactory (CNG2) channel have overall sequence identities of 60–70%. The calculated molecular masses range from 75–85 kD. The molecular mass of the rod photoreceptor channel α subunit deduced from the cDNA (79 kD) differs from the mass of the purified protein from rod outer segment (63 kD) estimated by SDS-PAGE. This difference can be attributed to the post-translational cleavage of a portion of the cytoplasmic N-terminus (Molday et al. 1991). There is evidence that the cone photoreceptor channel (CNG3) is also truncated in the N-terminal region (Bönigk et al. 1993; Weyand et al. 1994). A CNG channel that may represent a splice variant of the cone photoreceptor channel has been recently identified in taste buds of rat tongue (Misaka et al. 1997) suggesting that CNG channels are also implicated in taste chemoreception.

Although CNG channel β subunits (CNG4 and CNG5) are structurally related to CNG1–3 they fail to induce the formation of cGMP- or cAMP-gated channels. However, recent studies indicate that CNG5 may form a homomeric, NO-activated channel in neurons of the vomeronasal organ which is important for the sensory transduction of pheromones (Berghard et al. 1996; Broillet and Firestein 1997). CNG channel β subunits impart to the respective α subunits specific properties that are also found with native channels. For this reason, β subunits have been designated modulatory subunits of CNG channels. The two types of β subunits have been originally cloned from photoreceptors (CNG4) and olfactory neurons (CNG5), respectively. However, there is evidence that a specific CNG4 isoform is also expressed in olfactory neurons and may represent a subunit of the native olfactory channel (Sautter et al. 1998).

Hydropathicity analysis and sequence alignments indicate that CNG4 and CNG5, like the CNG channel α subunits, contain six transmembrane segments, an ion conducting pore and a putative binding pocket for cGMP in the C-terminus. Despite this general structural relationship, the overall sequence identity between α subunits and β subunits (30–50%) is significantly lower than the homology between the different α subunits (60–70%). In contrast to the α subunits and the CNG5 subunit, the primary transcript of the CNG4 gene is extensively spliced giving rise to several isoforms (Biel et al. 1996a) which are expressed in a tissue- and species-specific manner.

CNG channels have also been cloned from invertebrates like *D. melanogaster* (Baumann et al. 1994) and *C. elegans*. The tax-2 (Coburn and Bargmann 1996) and tax-4 (Komatsu et al. 1996) proteins of *C. elegans* are coexpressed in olfactory, gustatory and thermosensory neurons implicating that these subunits form a single heteromeric channel which is involved in multiple sensory modalities. The sequences of at least two additional members of the CNG channel family whose functions are still unknown have been detected in the genome of *C. elegans* (Fig. 1).

3
Channel Structure and Stoichiometry

Functionally, CNG channels belong to the class of ligand-gated cation channels since they are activated by the binding of a ligand (cGMP or cAMP) to the cyclic nucleotide-binding pocket. However, inspection of the primary sequences of the cloned channels revealed that CNG channels belong structurally to the superfamily of voltage-gated cation channels (Kaupp et al. 1989; Guy et al. 1991). The proposed structural model of the CNG channel is shown in Fig. 2. Like voltage-gated potassium channels CNG channels contain six transmembrane segments (S1–S6). The ion-conducting pore is

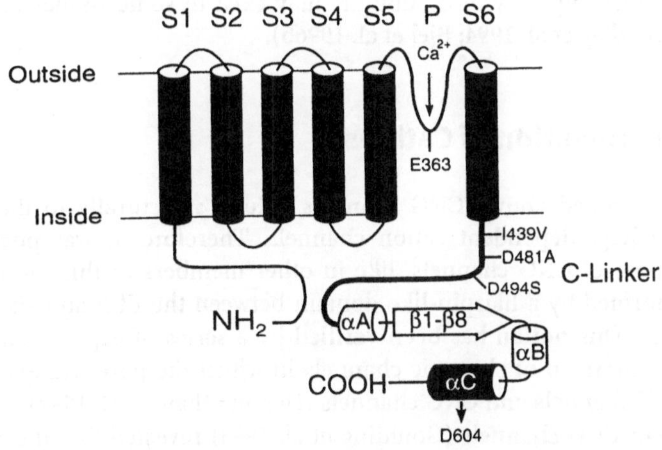

Fig. 2. Two dimensional model of the CNG channel. S1–S6, α helical transmembrane segments; P, pore region. The eight β strands (β1–β8) and the three α helices (αA-αC) that form the cyclic nucleotide-binding domain (CNBD) are schematically shown in the C-terminus of the channel. The C-linker which connects S6 with the CNBD is illustrated by a thick line. The three amino acids in the C-linker of CNG3 and CNG2 that determine cAMP efficacy, as well as the positions of D604 and E363 which are important for ligand selectivity and Ca^{2+} block, respectively, are indicated

formed by a "hairpin" structure between the S5 and S6 region (MacKinnon 1995). Both N- and C-termini are localized on the cytoplasmic site of the membrane. The presence of six transmembrane segments has been verified experimentally in the bovine CNG1 subunit by a gene fusion approach using the bacterial reporter enzymes alkaline phosphatase and β-galactosidase (Henn et al. 1995). All CNG channel subunits contain the positively charged "voltage sensor" motif in S4 (Kaupp et al. 1989; Goldstein 1996). However, the number of regularly spaced positive charges (3–4 basic residues at every third position) is reduced in comparison with the S4 segment of potassium channels (5–7 basic residues). This reduced number of positive residues may explain the very weak voltage-dependence of CNG channel opening (Zagotta and Siegelbaum 1996). A chimeric *Drosophila* ether-à-go-go K^+ channel containing the voltage sensor of the rat olfactory channel is still activated by voltage (Tang and Papazian 1997) indicating that the S4 segment of the CNG channel is functionally intact but may be inactivated by other CNG channel domains.

Recent results from the heterologous expression of tandem dimers of CNG channels demonstrate that CNG channels like the related voltage-gated potassium channels are tetrameric complexes (Gordon and Zagotta 1995a; Liu et al. 1996). Most native CNG channels may be composed of both α and β subunits. However, there is also evidence that homomeric CNG channels consisting only of α subunits may exist in some tissues (Torre et al. 1992; Bradley et al. 1994; Biel et al. 1996b).

4 Permeation of Cations

As stated above, CNG channels belong structurally to the superfamily of voltage-dependent cation channels. Therefore, it was postulated that the pore of CNG channels, like in other members of this ion channel class, is formed by a hairpin-like domain between the fifth and sixth segment (Fig. 2). This notion has been verified by a series of experimental findings. The expression of chimeric channels in which the pore was exchanged between K^+ channels and CNG channels (Heginbotham et al. 1992) or between different CNG channels (Goulding et al. 1993) revealed that the specific permeation properties of the channels mainly depend on the postulated pore regions. Both rod and olfactory CNG channels interact with the "ball peptide" that produces rapid inactivation in Shaker-type K^+ channels (Hoshi et al. 1990) by binding to a region within the pore (Kramer et al. 1994). Cysteine scanning mutagenesis indicated that the P region of the CNG channel like that of K^+ channels does not dip into and out of the membrane as initially proposed but extends toward the central axis of the channel forming the blades of an iris-like structure (Sun-et al. 1996). If one assumes a tetrameric channel, the S5 and S6 segments of the four subunits form the large-diameter internal vestibule of the channel, whereas the P regions may form the narrowest part of the pore that serves as the selectivity filter.

In the absence of divalents, CNG channels conduct a variety of monovalent cations like Na^+ and K^+ (Menini 1990; Frings et al. 1992; Haynes 1995a). By contrast, CNG channels preferentially conduct Ca^{2+} at physiological conditions i.e. in the presence of millimolar concentrations of extracellular Ca^{2+} (Fig. 3). Measurement of the permeability ratios between Ca^{2+} and Na^+ revealed that CNG channels are more permeable for Ca^{2+} than for Na^+. Especially the CNG3 channel (Weyand et al. 1994; Biel et al. 1994; Frings et al. 1995) and the CNG2 channel (Frings et al. 1995) strongly select Ca^{2+} over Na^+. Site directed mutagenesis has identified amino acid residues within the P region which determine the specific permeation properties. A tyrosine-glycine (YG)-motif, present in the pore of K^+ channels but not in that of CNG channels constitutes the selectivity filter for K^+ over Na^+ (Jan and Jan 1992; see also Doyle et al. 1998). When the YG-motif is deleted in K^+ channels, selectivity among monovalent cations is largely lost (Heginbotham et al. 1992). Ca^{2+} efficiently permeates CNG channels but also partially blocks

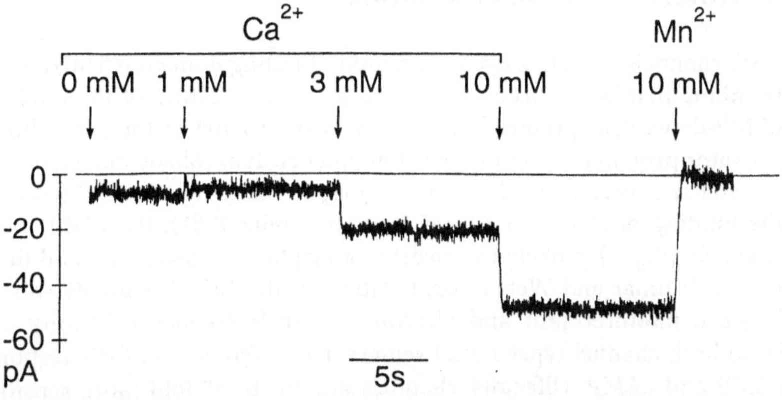

Fig. 3. Whole cell measurement of the calcium current through the heteromeric CNG3/CNG4 channel expressed in HEK 293 cells. The cell was clamped to 0 mV and dialysed with a pseudointracellular solution (140 mM K-aspartate, 8 mM NaCl, 10 mM EGTA, 10 mM HEPES pH 7.2) supplemented with 1 mM cGMP. After initial superfusion with a Ca^{2+} free Ringer's solution (140 mM NaCl, 5 mM KCl, 10 mM HEPES, 1 mM EGTA pH 7.2) the cell was subsequently superfused with Ringer's solutions supplemented with increasing concentrations of Ca^{2+}.The NaCl concentration was reduced according to the respective concentrations of Ca^{2+}, EGTA was omitted. The resulting inward current is carried by Ca^{2+} and is reversibly blocked when extracellular Ca^{2+} is exchanged by 10 mM Mn^{2+}

the channels in a voltage-dependent manner (Zufall and Firestein 1993; Haynes 1995b; Zagotta and Siegelbaum 1996). A negatively charged glutamate residue, adjacent to the YG-motif in the P region of CNG1–3 (E363 in CNG1; Fig. 2) is a major determinant of this block (Root and MacKinnon 1993; Eismann et al. 1994). The CNG4 subunit has a glycine residue replacing the glutamate residue, and this may explain why heteromeric photoreceptor channels containing the CNG4 subunit show less block by Ca^{2+} than homomeric CNG1–3 channels (Chen et al. 1993; Körschen et al. 1995).

There are striking similarities between the ion permeation properties of CNG channels and those of voltage-dependent calcium channels: 1. Like CNG channels, voltage-dependent Ca^{2+} channels conduct monovalent cations in the absence of divalent cations whereas the flow of monovalent cations is blocked in the presence of Ca^{2+} (Hess and Tsien 1984). 2. The pore of calcium channels also lacks the YG-motif and contains a ring of glutamate residues which determines the ion selectivity (Yang et al. 1993). 3. The permeation properties of both voltage-dependent calcium channels and CNG channels are consistent with a multi-ion occupancy of the pore region (Sesti et al. 1995).

5 Channel Activation and Gating

CNG channels contain a cyclic nucleotide-binding domain (CNBD) in the C-terminus that bears structural homology to the CNBDs of the cAMP- and cGMP-dependent protein kinases and to the CNBD of the catabolite gene activator protein (CAP) of *E. coli* (Kaupp et al. 1989; Shabb and Corbin 1992; Zagotta and Siegelbaum 1996). In analogy to the X-ray resolved structure of the binding domain of CAP (McKay and Steitz 1981), the CNBD of CNG channels (Fig. 2) is likely to consist of an eight-stranded β-roll and three α-helices (Kumar and Weber 1992). Although the CNBD is highly conserved between photoreceptor and olfactory channels (sequence identity of 80–90%) both channel types reveal remarkable differences in their response to cGMP and cAMP. Olfactory channels are 30- to 50-fold more sensitive to cGMP, and up to 1000-fold more sensitive to cAMP than photoreceptor channels (Zagotta and Siegelbaum 1996; Frings et al. 1992). In addition, cAMP acts as a partial agonist on photoreceptor channels, activating only a fraction of the current induced by cGMP, whereas it fully activates olfactory channels. Recent studies with mutated and chimeric CNG channels have elucidated key structural elements within the primary structure of the channel which determine these differences. The selectivity of CNG channels for cGMP is mainly determined by a negatively charged amino acid localized in the α C-helix of the CNBD (Goulding et al. 1994; Varnum et al. 1995). Substituting the respective aspartate residue in the α C-helix of CNG1 (D604; Fig. 2) by a nonpolar residue converts CNG1 from a channel that is highly selective for cGMP to a channel with greater cAMP sensitivity (Varnum et al. 1995). Interestingly, the molecular structures affecting the apparent affinity for cGMP or cAMP are not restricted to the CNBD but are distributed throughout the channel sequence. The N-terminal segment (Goulding et al. 1994; Gordon and Zagotta 1995b) and the linker peptide (C-linker) connecting the S6 transmembrane segment and the CNBD (Fig. 2) are especially influential in determining apparent agonist affinity and channel gating. The importance of the C-linker is emphasized by several findings. 1. The exchange of three amino acid residues in the C-linker of the cone photoreceptor channel by the respective residues of the olfactory channel (I439V, D481A and D494S; Fig. 2) dramatically increases the cAMP-efficacy of the cone channel (Zong et al. 1998). 2. Histidine residues have been identified in the C-linker of the CNG1 channel (H420) and the CNG2 channel (H396), that confer potentiation or inhibition of the corresponding currents by micromolar concentrations of Ni^{2+} (Gordon and Zagotta 1995b; 1995c). 3. An additional histidine which is present in all α subunits (H468 in bovine CNG1) has been found to be involved in channel gating (Gordon et al. 1996). 4. Donors of nitric oxide as well as other agents that modify free SH groups, activate native and expressed olfactory CNG channels in the absence of cGMP or cAMP presumably by covalently binding to a conserved cysteine residue which is also located in the C-linker (Broillet and Firestein 1996; 1997). 5. Mild oxidants like copper phenanthroline potentiate activation of the rod photoreceptor channel (Gordon et al. 1997) by facilitating the formation of a disulfide bond between two cysteine residues present in the C-linker (C481) and the N-terminus (C35).

In summary, CNG channel activation may require a concerted conformational change of various protein domains. Overlay assays suggest that the cytoplasmic N-terminal domain of CNG channels directly binds to domains of the C-terminus including the C-linker and part of the CNBD (Gordon et al. 1997; Varnum and Zagotta 1997). This interdomain interaction which may occur within the same subunit or between different channel subunits of the tetrameric complex is likely to constitute a crucial part of the channel gating machinery. Based on the observation that CNG channels reveal spontaneous openings in the absence of any cyclic nucleotide, an allosteric model of channel activation has been proposed (Tibbs et al. 1997). According to this model, cyclic nucleotides stabilize the channel in the open state thereby promoting a prominent increase of the open probability. However, it has been demonstrated recently that partially liganded CNG channels can move freely between up to three distinct subconductance states (Ruiz et al. 1997). This finding clearly contrasts with the simple closed-open transition postulated in cyclic allosteric models indicating that a more complex model will be necessary to precisely describe CNG channel activation.

6
CNG Channel β Subunits

Two types of CNG channel β subunits have been cloned, CNG4 and CNG5 (Fig. 1). Although the sequence identity is rather low between CNG4 and CNG5 (25%), both subunits reveal functional similarities. When coexpressed with the respective α subunit, they form heteromeric complexes and induce channel properties that are present in the native channels but are absent from the channels formed by the α subunits alone (Chen et al. 1993; Bradley et al. 1994; Liman and Buck 1994; Körschen et al. 1995; Biel et al. 1996b). The β subunit-induced properties include single-channel flickering, current

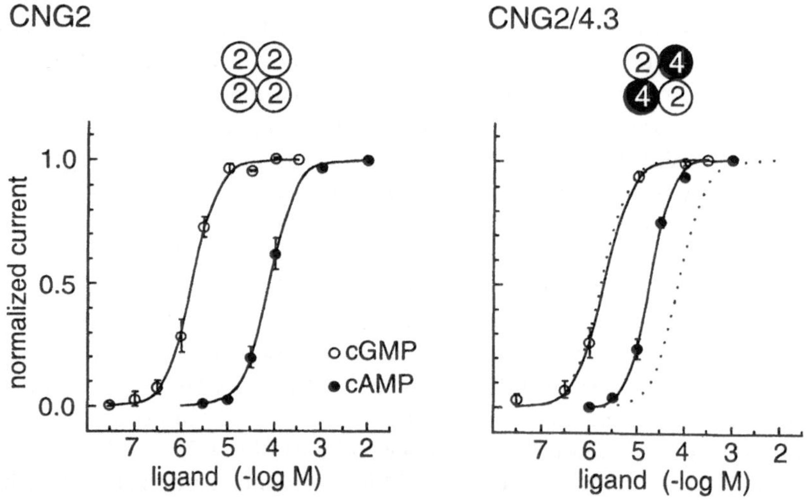

Fig. 4. Dose-response curves for cAMP and cGMP of the homomeric (CNG2) and heteromeric (CNG2/CNG4.3) olfactory channel measured at +60 mV. CNG4.3 represents the short CNG4 isoform that has been cloned from rat olfactory epithelium (Sautter et al. 1998; see also Fig. 5.) The 2:2 stoichiometry of CNG2 and CNG4.3 subunits in the heteromeric channel has not been experimentally proven and was choosen arbitrarily. To demonstrate the specific increase of the cAMP affinity induced by CNG4.3, the dose-response curves of the homomeric CNG2 channel have been included as dotted lines in the right diagram.The K_a value for cAMP is 74 μM for CNG2 and 18 μM for CNG2/CNG4.3, the K_a value for cGMP is about 2 μM for both channel types

blockade by L-cis diltiazem, weakening of the outward-rectification induced by external calcium and increase of the apparent affinity for cAMP. The latter effect seems to be especially important in the case of the olfactory channel (Fig. 4) since cAMP is the natural second messenger of olfaction (Firestein 1992). The participation of CNG4 in agonist binding was confirmed by photoaffinity labelling with an analogue of cGMP (Brown et al. 1995) that identified a peptide corresponding to the predicted CNBD of CNG4.

So far, only a single transcript coding for CNG5 has been identified. In contrast, the cloning of CNG4 yielded multiple transcripts that encode a whole family of isoforms (Biel et al. 1996a). The various CNG4 isoforms are derived from a complex gene locus by the use of multiple starting points for transcription and by alternative splicing of the primary transcript (Ardell et al. 1996; Sautter et al. 1998). The peptides differ from each other only in the length and the sequence of the cytoplasmic N-terminus (Fig. 5).

"Long" isoforms of CNG4 such as the subunits expressed in rod photoreceptors (Körschen et al. 1995; Ardell et al. 1996; Colville and Molday 1996) and in pineal gland (Sautter et al. 1997) are characterized by an extended N-terminus that is negatively charged due to the presence of a large number of glutamate residues. Interestingly, this glutamic-acid rich domain is also expressed as a distinct soluble protein (the glutamic acid-rich protein or GARP) which has been cloned from retina (Sugimoto et al. 1991; Colville and

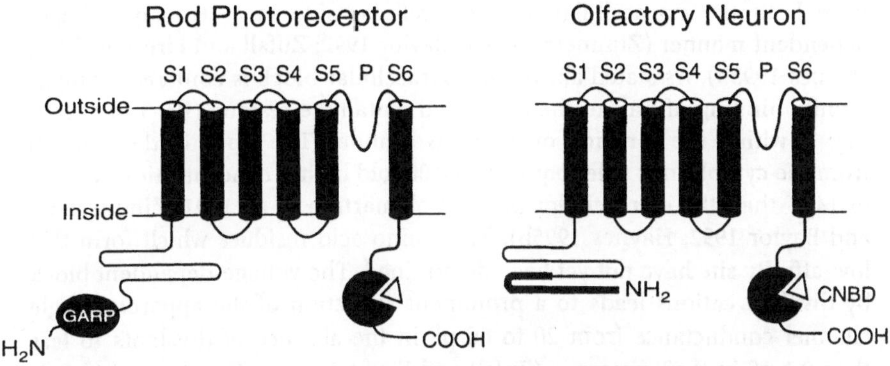

Fig. 5. Two-dimensional models of long and short CNG4 isoforms. The long isoform *(left)* is expressed in rod photoreceptors (Körschen et al. 1995) and pineal gland (Sautter et al. 1997), the short isoform (CNG4.3; *right)* has been cloned from olfactory neurons (Sautter et al. 1998). S1–S6, α helical transmembrane segments; P, pore region; CNBD, cyclic nucleotide-binding domain. The isoforms only differ from each other in their N-terminus that either contains or lacks the glutamic acid-rich protein (GARP) domain. The unique N-terminal segment of CNG4.3 is illustrated by a thick line

Molday 1996; Ardell et al. 1995) and pineal gland (Sautter et al. 1997). The physiological function of GARP and of the glutamic acid-rich domain present in the N-terminus of long CNG4 isoforms is not known. Comparison of the primary structures of different long CNG4 isoforms expressed in rat, bovine and human tissues revealed a surprisingly unequal distribution of sequence homology. The core region of CNG4 including the transmembrane domain and the CNBD is highly conserved between the different species (sequence identity of over 90%). However, the corresponding GARP-domains strongly diverge with homologies of only 50–60%. At present, there is no plausible explanation for such a large divergence among mammals in the "same" gene.

In "short" CNG4 subunits, the N-terminal GARP domain is replaced by various sequences depending on the respective isoform. Examples of short isoforms are the CNG4 subunits expressed in bovine testis (Biel et al. 1996b) and in olfactory epithelium (Sautter et al. 1998; Fig. 5). A short isoform may also be present in retina (Ardell et al. 1996).

7
Channel Modulation by Ca^{2+}

Recent studies have focused on the modulation of CNG channels by Ca^{2+}. Ca^{2+} influences the activity of CNG channels via different pathways. Ca^{2+} efficiently permeates CNG channels (Fig. 3) but also inhibits these channels from both the extracellular and intracellular side in a strongly voltage-dependent manner (Zimmerman and Baylor 1992; Zufall and Firestein 1993; Haynes 1995b). As stated above, the extracellular block is confered by high-affinity binding of Ca^{2+} to the conserved glutamate residue (E363 in CNG1; Fig. 1) within the P region of the CNG channel. The block of the channel from the cytoplasmic side requires 50–100 fold higher concentrations of Ca^{2+} or Mg^{2+} than the extracellular block (Colamartino et al. 1991; Zimmerman and Baylor 1992; Haynes 1995b). The amino acid residues which form this low-affinity site have not yet been determined. The voltage-dependent block by divalent cations leads to a prominent reduction of the apparent single channel conductance from 20 to 60 pS in the absence of divalents to less than 0.1 pS in their presence (Zufall and Firestein 1993; Zagotta and Siegelbaum 1996). At least for CNG channels expressed in sensory neurons the reduction of the single channel conductance by Ca^{2+} or Mg^{2+} is of crucial physiological significance because it improves the signal-to-noise ratio of sensory transduction. This is because the current generated through the gating of many low conductance channels will be less noisy than the current induced by a few high conductance channels.

In addition to its direct interaction with CNG channels, Ca^{2+} causes a decrease in the apparent cyclic nucleotide affinity of both photoreceptor and olfactory channels by activation of calmodulin (Hsu and Molday 1993; Chen and Yau 1994; Gordon et al. 1995). This inhibitory action of Ca^{2+}-calmodulin (CaM) is a key component of olfactory adaptation (Chen and Yau 1994; Kurahashi and Menini 1997) and may also contribute to light adaptation (Hsu and Molday 1993), although to a lesser degree. The modulation by CaM does not involve the activation of a kinase but is mediated by the direct binding of CaM to the channel. The CaM modulation of the native rod photoreceptor channel is confered by the β subunit (Chen et al. 1994; Körschen et al. 1995; Grunwald et al. 1998; Weitz et al. 1998). Two distinct CaM binding sites have been identified in the primary sequence of the CNG4 subunit (Grunwald et al. 1998; Weitz et al. 1998). These sites are localized in the cytoplasmic N-terminus preceding the first transmembrane segment and in the C-terminus downstream of the CNBD, respectively. Interestingly, coexpression of CNG1/CNG4 heteromers revealed that deletion of the N-terminal CaM binding side abolishes current inhibition by CaM, whereas deletion of the C-terminal side has no major effect on current modulation by CaM. In contrast to the rod channel, CaM controls the activity of the olfactory channel (Liu et al. 1994) by binding to a single site situated on the CNG2 α subunit. The inhibition of the olfactory channel by CaM is much more pronounced (up to 50 fold shift of K_a for cGMP) than that found with rod and cone photoreceptor channels (2–3 fold shift of K_a). The CaM binding site of the CNG2 subunit has been mapped to its N-terminus (Liu et al. 1994). It is formed by a 26-amino acid stretch that, unlike both CaM binding sites identified in the CNG4 subunit, corresponds to the consensus sequence of CaM binding sites found in other CaM binding proteins (Ikura et al. 1992). Recently, it has been shown that the binding site for CaM constitutes an autoexcitatory channel domain that binds to the C-terminus of the CNG channel (Varnum and Zagotta 1997). Thus, CaM may inhibit channel activity by targeting this interdomain interaction.

Ca^{2+} is likely to modulate CNG channels also by interaction with cellular proteins that are different from calmodulin. There is evidence for the presence of a Ca^{2+} binding protein that induces calmodulin-like inhibition of the frog rod photoreceptor channel (Gordon et al. 1995). This protein has apparently different Ca^{2+} dependence and/or higher affinity for the channel than calmodulin.

8
CNG Channels in Non-Sensory Tissues

The cDNAs of CNG channel subunits have been cloned from various non-sensory tissues (Finn et al. 1996; Biel et al. 1996a). The high permeability of CNG channels for Ca^{2+} suggests that CNG channels function in these tissues as cyclic nucleotide-operated calcium channels which transduce a hormonally induced rise in the intracellular cGMP/cAMP-level into an influx of Ca^{2+}. Such a mechanism has been identified in chick (Dryer and Henderson 1991) and rat (Schaad et al. 1995) pinealocytes where CNG channels may be involved in the regulation of the synthesis and secretion of melatonin. PCR analysis and cloning revealed that chick pineal expresses both rod photoreceptor (CNG1) and cone photoreceptor (CNG3) α subunits (Bönigk et al. 1996). Similarly, a rod photoreceptor CNG channel consisting of the CNG1 α subunit and a long CNG4 β subunit has been cloned from rat pineal (Sautter et al. 1997). Photoreceptor-type channels have also been detected in retinal On-bipolar cells (Shiells and Falk 1990; Nawy and Jahr 1991), retinal ganglion cells (Ahmad et al. 1994) and sympathetic neurons (Thompson 1997). CNG channels may be functionally important for development and normal function of the mammalian brain (Zufall et al. 1997). Both olfactory and rod photoreceptor channels have been detected in CA1–3 hippocampal neurons (Kingston et al. 1996; Bradley et al. 1997). It has been speculated that CNG channels regulate neurotransmitter release in response to the retrograde messenger NO, that potentiates synaptic transmission in the brain (Zufall et al. 1997). The importance of CNG channels for modulation of synaptic strength is highlighted by a recent report demonstrating that a CNG channel mediates synaptic feedback by nitric oxide in synapses between cone photoreceptors and horizontal cells of the retina (Savchenko et al. 1997).

CNG channels are not restricted to the central nervous system but are also found in various peripheral tissues. A CNG channel has been found in mammalian sperm cells (Weyand et al. 1994). The physiological role of the channel is not known yet, however influx of Ca^{2+} through a CNG channel may constitute a crucial part of the machinery that controls chemotactic movement of sperm cells (Garbers 1989). There is a good deal of uncertainty about the physiological relevance of CNG channels in other organs like kidney, heart and aorta (Distler et al. 1994; Ding et al. 1997). In most of these tissues cGMP- or cAMP-activated currents have not be detected *in vivo* up to now. There are at least two explanations for this finding. 1. The low density of CNG channels in the cell may be necessary to prevent a Ca^{2+} overflow when the channel is activated since CNG channels, unlike many other cation channels, do not desensitize in the presence of an agonist. 2. CNG channels may be only expressed in specific subsets of cells within a given organ.

9 Concluding Remarks

The past few years have witnessed a dramatic increase in our understanding of the structure and function of CNG channels. Whereas it was initially assumed that CNG channels are specifically expressed in sensory neurons of the retina and the olfactory epithelium, it is now clear that these channels are distributed throughout the body. Molecular cloning revealed that CNG channels comprise a multi-gene family of at least five members. In addition, there is evidence for the presence of a whole variety of cyclic nucleotide-modulated channels whose primary structure is not known in most cases. It is now firmly established that the major function of CNG channels consists in providing a second messenger-triggered influx pathway for Ca^{2+}. Depending on the cell type, a rise of intracellular Ca^{2+} could trigger such diverse functions as cell motility, secretion and neuronal plasticity.

Acknowledgements. The research conducted in the author's laboratory was supported by grants from Fonds der Chemie, BMBF and Deutsche Forschungsgemeinschaft.

References

Ahmad I, Leinders-Zufall T, Kocsis JD, Shepherd GM, Zufall F, Barnstable CJ (1994). Retinal ganglion cells express a cGMP-gated cation conductance activatable by nitric oxide donors. Neuron 12:155–166

Arancio O, Kandel ER, Hawkins RD (1995) Activity-dependent long-term enhancement of transmitter release by presynaptic 3',5'-cyclic GMP in cultured hippocampal neurons. Nature 376:74–80

Ardell MD, Aragon I, Oliveira L, Porche GE, Burke E, Pittler SJ (1996) The β subunit of human rod photoreceptor cGMP-gated cation channel is generated from a complex transcription unit. FEBS Lett 389:213–218

Ardell MD, Makhija AK, Olivera L, Miniou P, Viegas-Péquignot E, Pittler SJ (1995) cDNA, gene structure, and chromosomal localisation of human GAR1 (CNCG3L), a homolog of the third subunit of bovine photoreceptor cGMP-gated channel. Genomics 28:32–38

Baumann A, Frings S, Godde M, Seifert R, Kaupp UB (1994) Primary structure and functional expression of a Drosophila cyclic nucleotide-gated channel present in eyes and antennae. EMBO J 13:5040–5050

Baylor D (1996) How photons start vision Proc Natl Acad Sci USA 93:560–565

Beavo JA (1995) Cyclic nucleotide phosphodiesterases: functional implications of multiple isoforms. Physiol Rev 75:725–748

Berghard A, Buck LB, Liman ER (1996) Evidence for distinct signaling mechanism in two mammalian olfactory sense organs. Proc Natl Acad Sci USA 93:2365–2369

Biel M, Zong X, Distler M, Bosse E, Klugbauer N, Murakami M, Flockerzi V, Hofmann F (1994) Another member of the cyclic nucleotide-gated channels family expressed in testis, kidney and heart. Proc Natl Acad Sci USA 91:3505–3509

Biel M, Zong X, Hofmann F (1996a) Cyclic nucleotide-gated cation channels: molecular diversity, structure and cellular functions. Trends Cardiovasc Med 6:274–280

Biel M, Zong X, Ludwig A, Sautter A, Hofmann F (1996b) Molecular cloning and expression of a modulatory subunit of the cyclic nucleotide-gated cation channel. J Biol Chem 271:6349–6355

Bönigk W, Altenhofen W, Müller F, Dose A, Illing M, Molday RS, Kaupp UB (1993) Rod and cone photoreceptor cells express distinct genes for cGMP-gated channels. Neuron 10:865–877

Bönigk W, Müller F, Middendorff R, Weyand I, Kaupp UB (1996) Two alternatively spliced forms of the cGMP-gated channel α-subunit from cone photoreceptor are expressed in the chick pineal organ. J Neuroscience 16:7458–7468

Bradley J, Li J, Davidson N, Lester HA, Zinn K (1994) Heteromeric olfactory cyclic nucleotide-gated channels: A new subunit that confers increased sensitivity to cAMP. Proc Natl Acad Sci USA 91:8890–8894

Bradley J, Zhang Y, Bakin R, Lester HA, Ronnett GV, Zinn K (1997) Functional expression of the heteromeric "olfactory" cyclic nucleotide-gated channel in hippocampus: a potential effector of synaptic plasticity in brain neurons. J Neuroscience 17:1993–2005

Breer H, Raming K, Krieger J (1994) Signal recognition and transduction in olfactory neurons. Biochim Biophys Acta 1224:277–287

Broillet MC, Firestein S (1996) Direct activation of the olfactory cyclic nucleotide-gated channel through modification of sulfhydryl groups by NO compounds. Neuron 16:377–385

Broillet MC, Firestein S (1997) β subunits of the olfactory cyclic nucelotide-gated channel form a nitric oxide activated Ca^{2+} channel. Neuron 18:951–958

Brown RL, Gramling R, Bert RJ, Karpen JW (1995) Cyclic GMP contact points within the 63-kDa subunit and a 240-kDa associated protein of retinal rod cGMP-activated channels. Biochemistry 34:8365–8370

Brunet LJ, Gold GH, Ngai J (1996) General anosmia caused by a targeted disruption of the mouse olfactory cyclic nucleotide-gated cation channel. Neuron 17:681–693

Chen TY, Illing M, Hsu YT, Yau KW, Molday RS (1994) Subunit 2 (or β) of retinal rod cGMP-gated cation channel is a component of the 240-kDa channel-associated protein and mediates Ca^{2+}-calmodulin modulation. Proc Natl Acad Sci USA 91:11757–11761

Chen TY, Peng YW, Dhallan RS, Ahamed B, Reed RR, Yau KW (1993) A new subunit of the cyclic nucleotide-gated cation channel in retinal rods. Nature 362:764–767

Chen TY, Yau KW (1994) Direct modulation by Ca^{2+}-calmodulin of cyclic nucleotide-activated channel of rat olfactory receptor neurons. Nature 368:545–548

Cobbs WH, Barkdoll AE III, Pugh EN Jr (1985) Cyclic GMP increases photocurrent and light sensitivity of retinal cones. Nature 317:64–66

Coburn CM, Bargmann CI (1996) A putative cyclic nucleotide-gated channel is required for sensory development and function in C. elegans. Neuron 17:695–706

Colamartino G, Menini A, Torre V (1991) Blockage and permeation of divalent cations through the cyclic GMP-activated channel from tiger salamander retinal rods. J Physiol 440:189–206

Colville CA, Molday RS (1996) Primary structure and expression of the human β-subunit and related proteins of the rod photoreceptor cGMP-gated channel. J Biol Chem 271:32968–32974

Cook NJ, Hanke W, Kaupp UB (1987) Identification, purification, and functional reconstitution of the cyclic GMP-dependent channels from rod photoreceptors. Proc Natl Acad Sci USA 84:585–589

Dhallan RS, Yau KW, Schrader KA, Reed RR (1990) Primary structure and functional expression of a cyclic nucleotide-activated channel from olfactory neurons. Nature 347:184–187

Ding C, Potter ED, Qiu W, Coon SL, Levine MA, Guggino SE (1997) Cloning and widespread distribution of the rat rod-type cyclic nucleotide-gated cation channel. Am J Physiol 272:C1335–1344

Distler M, Biel M, Flockerzi V, Hofmann F (1994) Expression of cyclic nucleotide-gated cation channels in non-sensory tissues and cells. Neuropharmacology 33:1275–1282

Doyle DA, Cabral JM, Pfuetzner RA, Kuo A, Gulbis J, Cohen SL, Chait BT, MacKinnon R (1998) The structure of the potassium channel: molecular basis of K^+ conduction and selectivity. Science 280:69–77

Dryer S, Henderson D (1991) A cyclic GMP-activated channel in dissociated cells of chick pineal gland. Nature 353:756–758

Eismann E, Müller F, Heinemann SH, Kaupp UB (1994) A single negative charge within the pore region of a cGMP-gated channel controls rectification, Ca^{2+} blockage, and ion selectivity. Proc Natl Acad Sci USA 91:1109–1113

Fesenko EE, Kolesnikov SS, Lyubarsky AL (1985) Induction by cyclic GMP of cationic conductance in plasma membrane of retinal rod outer segments. Nature 313:310–313

Finn JT, Grunwald ME, Yau KW (1996) Cyclic nucleotide-gated ion channels: an extanded family with diverse functions. Annu Rev Physiol 58:395–426

Finn JT, Solessio EC, Yau KW (1997) A cGMP-gated cation channel in depolarizing photoreceptors of the lizard parietal eye. Nature 385:815–819

Firestein S (1992) Electric signals in olfactory transduction. Current Opinion in Neurobiology 2:444–448

Frings S, Lynch JW, Lindemann B (1992) Properties of cyclic nucleotide-gated channels mediating olfactory transduction: Activation, selectivity, and blockage. J Gen Physiol 100:45–67

Frings S, Seifert R, Godde M, Kaupp UB (1995) Profoundly different calcium permeation and blockage determine the specific function of distinct cyclic nucleotide-gated channels. Neuron 15:169–179

Garbers DL (1989) Molecular basis of fertilization. Annu Rev Biochem 58:719–742

Goldstein SAN (1996) A structural vignette common to voltage sensors and conduction pores: canaliculi. Neuron 16:717–722

Gordon SE, Downing-Park J, Zimmerman AL (1995) Modulation of the cGMP-gated ion channel in frog rods by calmodulin and an endogenous inhibitory factor. J Physiol 486:533–546

Gordon SE, Oakley JC, Varnum MD, Zagotta WN (1996) Altered ligand specificity by protonation in the ligand binding domain of cyclic nucleotide-gated channels. Biochemistry 35:3994–4001

Gordon SE, Varnum MD, Zagotta WN (1997) Direct interaction between amino- and carboxyl-terminal domains of cyclic nucleotide-gated channels. Neuron 19:431–441

Gordon SE, Zagotta WN (1995a). Subunit interactions in coordination of Ni^{2+} in cyclic nucleotide-gated channels. Proc Natl Acad Sci USA 92:10222–10226

Gordon SE, Zagotta WN (1995b) Localization of regions affecting an allosteric transition in cyclic nucleotide-activated channels. Neuron 14:857–864

Gordon SE, Zagotta WN (1995c) A histidine residue associated with the gate of the cyclic nucleotide-activated channels in rod photoreceptors. Neuron 14:177–183

Goulding EH, Tibbs GR, Liu D, Siegelbaum SA (1993) Role of H5 domain in determining pore diameter and ion permeation through cyclic nucleotide-gated channels. Nature 364:61–64

Goulding EH, Tibbs GR, Siegelbaum SA (1994) Molecular mechanism of cyclic nucleotide-gated channel activation. Nature 372:369–374

Grunwald ME, Yu WP, Yu HH, Yau KW (1998) Identification of a domain on the β-subunit of the rod cGMP-gated cation channel that mediates inhibition by calcium-calmodulin. J Biol Chem 273:9148–9157

Guy HR, Durell SR, Warmke J, Drysdale R, Ganetzky B (1991) Similarities in amino acid sequences of Drosophila eag and cyclic nucleotide-gated channels. Science 254:730

Hatt H, Ache BW (1994) Cyclic nucleotide- and inositol phosphate-gated ion channels in lobster olfactory neurons. Proc Natl Acad Sci USA 91:6264–6268

Haynes LW (1995a) Permeation of internal and external monovalent cations through the catfish cone photoreceptor cGMP-gated channel. J Gen Physiol 106:485–505

Haynes LW (1995b) Permeation and block by internal and external cations of the catfish cone photoreceptor cGMP-gated channel. J Gen Physiol 106:507–523

Haynes LW, Yau KW (1985) Cyclic GMP-sensitive conductance in outer segment membrane of the catfish cones. Nature 317:61–64

Heginbotham L, Abramson T, MacKinnon R (1992) A functional connection between the pores of distantly related ion channels as revealed by mutant K^+ channels. Science 258:1152–1155

Henn DK, Baumann A, Kaupp UB (1995) Probing the transmembrane topology of cyclic nucleotide-gated ion channels with a gene fusion approach. Proc Natl Acad Sci USA 92:7425–7429

Hess P, Tsien RW (1984) Mechanism of ion permeation through calcium channels. Nature 309: 453–456

Hofmann F, Dostmann W, Keilbach A, Landgraf W, Ruth P (1992) Structure and physiological role of cGMP-dependent protein kinase. Biochem Biophys Acta 1135:51–60

Hofmann F, Ludwig A, Pfeifer A (1994) Cyclic GMP and the control of airways smooth muscle tone. In: Raeburn D, Giembycz MA (eds) Airways smooth muscle: Biochemical control of contraction and relaxation. Birkhäuser Verlag, Basel, pp253–269

Hoshi T, Zagotta WN, Aldrich RW (1990) Biophysical and molecular mechanism of Shaker potassium channel inactivation. Science 250:533–538

Hsu YT, Molday RS (1993) Modulation of the cGMP-gated channel of rod photoreceptor cells by calmodulin. Nature 361:76–79

Ikura M, Clore GM, Gronenborn AM, Zhu G, Klee CB, Bax A (1992) Solution structure of a calmodulin-target peptide complex by multidimensional NMR. Science 256:632–638

Jan LY, Jan YN (1992) Structural elements involved in specific K^+ channel functions. Annu Rev Physiol 54:537–555

Kaupp UB, Niidome T, Tanabe T, Terada S, Bönigk W, Stühmer W, Cook NJ, Kangawa K, Matsuo H, Hirose T, Miyata T, Numa S (1989) Primary structure and functional expression from complementary DNA of the rod photoreceptor cyclic GMP-gated channel. Nature 342:762–766

Kingston PA, Zufall F, Barnstable CJ (1996) Rat hippocampal neurons express genes for both rod retinal and olfactory cyclic nucleotide-gated channels: novel targets for cAMP/cGMP function. Proc Natl Acad Sci USA 93:10440–10445

Komatsu H, Mori I, Rhee JS, Akaike N, Ohshima Y (1996) Mutations in a cyclic nucleotide-gated channel lead to abnormal thermosensation and chemosensation in C. elegans. Neuron 17:707–718

Körschen HG, Illing M, Seifert R, Sesti F, Williams A, Gotzes S, Colville C, Müller F, Dose A, Godde M, Molday L, Kaupp UB, Molday RS (1995) A 240 kDa protein represents the complete β subunit of the cyclic nucleotide-gated channel from rod photoreceptor. Neuron 15:627–636

Kramer RH, Goulding E, Siegelbaum SA (1994) Potassium channel inactivation peptide blocks cyclic nucleotide-gated channels by binding to the conserved pore domain. Neuron 12:655–662

Kumar VD, Weber IT (1992) Molecular model of the cyclic GMP-binding domain of the cyclic GMP-gated ion channel. Biochemistry 31:4643–4649

Kurahashi T, Menini A (1997) Mechanism of odorant adaptation in the olfactory receptor cell. Nature 385:725–729

Liman ER, Buck LB (1994) A second subunit of the olfactory cyclic nucleotide-gated channel confers high sensitivity to cAMP. Neuron 13:611–621

Liu DT, Tibbs GR, Siegelbaum SA (1996) Subunit stoichiometry of cyclic nucleotide-gated channels and effects of subunit order on channel function. Neuron 16:983–990

Liu M, Chen TY, Ahamed B, Li J, Yau KW (1994) Calcium-calmodulin modulation of the olfactory cyclic nucleotide-gated cation channel. Science 266:1348–1354

MacKinnon R (1995) Pore loops: an emerging theme in ion channel structure. Neuron 14:889–892

McKay DB, Steitz TA (1981) Structure of catabolite gene activator protein at 2.9 A resolution suggests binding to left-handed B-DNA. Nature 290:744–749

Menini A (1990) Currents carried by monovalent cations through cyclic GMP-activated channels in excised patches from salamander rods. J Physiol 424:167–185

Misaka T, Kusakabe Y, Emori Y, Gonoi T, Arai S, Abe K (1997) Taste buds have a cyclic nucleotide-activated channel, CNGgust. J Biol Chem 272:22623–22629

Molday RS, Molday LL, Dose A, Clark-Lewis I, Illing M, Cook NJ, Eismann E, Kaupp UB (1991) The cGMP-gated channel of the rod photoreceptor cell: Characterization and orientation of the amino terminus. J Biol Chem 266:21917–21922

Nakamura T, Gold GH (1987) A cyclic nucleotide-gated conductance in olfactory receptor cilia. Nature 325:442–444

Nawy S, Jahr CE (1991) cGMP-gated conductance in retinal bipolar cells is suppressed by the photoreceptor transmitter. Neuron 7:677–683

Root MJ, MacKinnon R (1993) Identification of an external divalent cation-binding site in the the pore of a cGMP-activated channel. Neuron 11:459–466

Ruiz ML, Karpen JW (1997) Single cyclic nucleotide-gated channels locked in different ligand-bound states. Nature 389:389–392

Sautter A, Biel M, Hofmann F (1997) Molecular cloning of cyclic nucleotide-gated cation channel subunits from rat pineal gland. Mol Brain Res 48:171–175

Sautter A, Zong X, Hofmann F, Biel M (1998). An isoform of the rod photoreceptor cyclic nucleotide-gated channel β subunit expressed in olfactory neurons. Proc Natl Acad USA 95:4696–4701

Savchenko A, Barnes S, Kramer RH (1997) Cyclic-nucleotide-gated channels mediate synaptic feedback by nitric oxide. Nature 390:694–698

Schaad NC, Vanecek J, Rodriguez IR, Klein DC, Holtzclaw L, Russel JT (1995) Vasoactive intestinal peptide elevates pinealocyte intracellular calcium concentrations by enhancing influx: evidence for involvement of a cyclic GMP-dependent mechanism. Mol Pharmacol 47:923–933

Sesti F, Eismann E, Kaupp UB, Nizzari M, Torre V (1995) The multi-ion nature of the cGMP-gated channel from vertebrate rods. J Physiol 487:17–36

Shabb JB, Corbin JD (1992) Cyclic nucleotide-binding domains in proteins having diverse functions. J Biol Chem 267:5723–5726

Shiells RA, Falk G (1990) Glutamate receptors of rod bipolar cells are linked to a cyclic GMP cascade via a G-protein. Proc R Soc Lond 242:91–94

Sugimoto Y, Yatsunami K, Tsujimoto M, Khorana HG, Ichikawa A (1991) The amino acid sequence of a glutamic acid-rich protein from bovine retina as deduced from the cDNA sequence. Proc Natl Acad Sci USA 88: 3116–3119

Sun ZP, Akabas MH, Goulding EH, Karlin A, Siegelbaum SA (1996) Exposure of residues in the cyclic nucleotide-gated channel pore: P region structure and function in gating. Neuron 16: 141–149

Tang CY, Papazian DM (1997) Transfer of voltage independence from a rat olfactory channel to the *Drosophila* ether-à-go-go K^+-channel. J Gen Physiol 109:301–311

Thompson SH (1997) Cyclic GMP-gated channels in a sympathetic neuron cell line. J Gen Physiol 110:155–164

Tibbs GR, Goulding EH, Siegelbaum SA (1997) Allosteric activation and tuning of ligand efficacy in cyclic nucleotide-gated channels. Nature 86:612–615

Torre V, Straforini M, Sesti F, Lamb TD (1992) Different channel-gating properties of two classes of cyclic GMP-activated channel in vertebrate photoreceptors. Proc R Soc London Ser B 250:209–215

Vaandrager AB, De Jonge HR (1994) Effect of cyclic GMP on intestinal transport. Adv Pharmacol 26:252–282

Varnum MD, Black KD, Zagotta WN (1995) Molecular mechanism for ligand discrimination of cyclic nucleotide-gated channels. Neuron 15:619–625

Varnum MD, Zagotta WN (1997) Interdomain interactions underlying activation of cyclic nucleotide-gated channels. Science 278:110–113

Weitz D, Zoche M, Müller F, Beyermann M, Körschen HG, Kaupp UB, Koch KW (1998) Calmodulin controls the rod photoreceptor CNG channel through an unconventional binding site in the N-terminus of the β-subunit. EMBO J 17:2273–2284

Weyand I, Godde M, Frings S, Weiner J, Müller F, Altenhofen W, Hatt H, Kaupp UB (1994) Cloning and functional expression of a cyclic nucleotide-gated channel from mammalian sperm. Nature 368:859–863

Yang J, Ellinor PT, Sather WA, Zhang JF, Tsien RW (1993) Molecular determinants of Ca^{2+} selectivity and ion permeation of L-type Ca^{2+} channels. Nature 366:158–161

Yau KW, Nakatani K (1985) Light-suppressible, cyclic GMP-sensitive conductance in the plasma membrane of a truncated rod outer segment. Nature 317:252–255

Zagotta WN, Siegelbaum SA (1996) Structure and function of cyclic nucleotide-gated channels. Annu Rev Neurosci 19:235–263

Zimmerman AL, Baylor DA (1992) Cation interactions within the cyclic GMP-activated channel of retinal rods from the tiger salamander. J Physiol 449:759–783

Zong X, Zucker H, Hofmann F, Biel M (1998) Three amino acids in the C-linker are major determinants of gating in cyclic nucleotide-gated channels. EMBO J 17:353–362

Zufall F, Firestein S (1993) Divalent cations block the cyclic nucleotide-gated channel of olfactory receptor neurons. J Neurophysiol 69:1758–1768

Zufall F, Firestein S, Shepherd GM (1994) Cyclic nucleotide-gated ion channels and sensory transduction in olfactory receptor neurons. Annu Rev Biophys Biomol Struct 23:577–607

Zufall F, Shepherd GM, Barnstable CJ (1997) Cyclic nucleotide-gated channels as regulator of CNS development and plasticity. Curr Opin Neurobiol 7:404–412

Signal Transduction by cGMP-Dependent Protein Kinases and Their Emerging Roles in the Regulation of Cell Adhesion and Gene Expression

M. Eigenthaler, S. M. Lohmann, U. Walter, and R. B. Pilz[*]

Institut für Klinische Biochemie und Pathobiochemie,
Medizinische Universitätsklinik, Josef-Schneider-Straße 2,
D-97080 Würzburg, Germany
and
[*] Department of Medicine, University of California, San Diego, La Jolla,
CA 92093-0652, USA

Contents

Abbreviations

ANP atrial natriuretic peptide; BHK baby hamster kidney cells; BNP brain natriuretic peptide; CFTR cystic fibrosis transmembrane conductance regulator; cAMP-PK cAMP-dependent protein kinase; cGMP-PK cGMP-dependent protein kinase; CNP natriuretic peptide type C; EC endothelial cell; EDRF endothelium-derived relaxing factor; GC guanylyl cyclase; GnRH gonadotropin releasing hormone; MKP-1 mitogen-activated protein kinase phosphatase; MLC myosin light chain; MLCK myosin light chain kinase; NLS nuclear localization signal; NO nitric oxide; NOS nitric oxide synthase; HUVEC human umbilical vein endothelial cell; 8-Br-cGMP 8-bromo-cGMP; PDGF platelet-derived growth factor; SHR spontaneously hypertensive rats; SMC smooth muscle cell; SNAP S-nitrosopenicillamine; SNP sodium nitroprusside; TGFβ1 tumor growth factor beta1; TNF tumor necrosis factor; VASP vasodilator-stimulated phosphoprotein; VEGF vascular endothelial growth factor;

1 Introduction

cGMP-dependent protein kinases (cGMP-PKs) have emerged as important signal transduction mediators of the effects of certain hormones, inter-/intracellular signals, toxins and drugs (Corbin and Lincoln 1978; Walter 1984 and 1989; Hofmann et al. 1992; Butt et al. 1993; Francis and Corbin 1994; Lincoln et al. 1995a; Vaandrager and DeJonge 1996; Lohmann et al. 1997). It is now widely recognized that a variety of peptide hormones [atrial natriuretic peptide (ANP), brain natriuretic peptide (BNP), natriuretic peptide type C (CNP), as well as enterotoxins and their endogenous homolog, guanylin] activate transmembrane receptor guanylyl cyclases (GC-A, GC-B, GC-C) and increase the cellular level of cGMP in many tissues and cell types. Many other hormones, cytokines and drugs cause the production and/or release of nitric oxide (NO) which then activates the heme-containing soluble guanylyl cyclases (sGC) and increases cGMP (Koesling et al. 1991; Schmidt and Walter 1994; Wedel and Garbers 1997). The discovery of at least 29 genes encoding putative GCs in the model system *Caenorhabditis elegans* perhaps foretells the numerous GCs and cGMP generating systems yet to be discovered in mammalian cells (Wedel and Garbers 1997).

The complexity of the cGMP-generating system is mirrored by an equally complex intracellular cGMP effector system. Established targets of cGMP include the protein families of cGMP-PKs, cGMP-stimulated/cGMP-inhibited phosphodiesterases, cGMP-gated ion channels, and possibly also cAMP-dependent protein kinases (cAMP-PKs) (Walter 1984 and 1989; Kaupp 1995; Rybalkin and Beavo 1996; Lohmann et al. 1997). A major challenge in this area of research is therefore to define which of the physiological effects of cGMP-elevating agents (i.e. NO, natriuretic peptides and others) are mediated by cGMP-PKs. The cGMP-dependent protein kinase, originally discovered using extracts of lobster tail muscle (Kuo and Greengard 1970), was first characterized and purified from bovine lung extracts (Gill et al. 1976; Lincoln et al. 1977). Subsequently, the complete mammalian amino acid sequence was elucidated by protein sequencing (Takio et al. 1984) and cDNA cloning (Sandberg et al. 1989; Wernet et al. 1989). These and additional data, particularly the recent characterization of the human gene (Orstavik et al. 1997) also established that the soluble cGMP-dependent protein kinase (type I) exists in two splice variants (type Iα and Iß). In *Drosophila*, two genes (DG1 and DG2) coding for cGMP-dependent protein kinases were identified and cloned (Kalderon and Rubin 1989). Interestingly, the recent assignment of mutations in the *foraging gene (for)* to the DG2 locus, and therefore the cGMP-dependent protein kinase, implicates the cGMP signal transduction pathway in the regulation of food-search behavior in *D. melanogaster* (Osborne et al. 1997). In mammalian cells, a distinct and primarily membrane-bound cGMP-dependent protein kinase (type II) was originally identified in and subsequently cloned from epithelial cells of the small intestine (DeJonge 1981; Jarchau et al. 1994). This type II cGMP-PK was also detected in other tissues including mouse brain (Uhler 1993) and rat kidney (Jarchau et al. 1994, Gambaryan et al. 1996). The membrane targeting of cGMP-PK II is required for some of its biological activities and is, at least in part, mediated by N-terminal myristoylation of the protein (Vaandrager et al. 1998).

This article will not discuss the structural and biochemical properties of cGMP-PK I and II since these have been extensively reviewed (Francis and Corbin 1994; Vaandrager and DeJonge 1996; Lohmann et al. 1997). Instead, some experimental and biochemical criteria will be presented which should be met for the conclusion that a given physiological/pharmacological effect is dependent on cGMP-PK activity. Some examples for which most of these criteria for cGMP-PK function were fulfilled will be emphasized. The emerging role of cGMP/cGMP-PK in the regulation of cell adhesion/migration and gene expression will be presented in subsections 3 and 4, respectively.

2
Cellular Distribution and Some Functional Roles of cGMP-PKs

Some of the present knowledge with respect to the expression and functional roles of cGMP-PKs is summarized in Table 1. However, the physiological function of the cGMP-PK has been firmly established only in some cases, since in most cases, the precise molecular function and mechanism of cGMP-PK action (which includes the identification of cGMP-PK substrates and their cellular roles) has not been elucidated. The conclusion that a given physiological/pharmacological effect of cGMP-elevating agents is mediated by the cGMP-PKs requires several lines of experimental evidence:

- NO, ANP or the factor being investigated should be able to increase cGMP in the target cell type or tissues,
- these cells/tissues should contain at least one form of cGMP-PK which is activated in response to cGMP-elevating agents in the intact cell system under investigation,
- specific activators/inhibitors of cGMP-PKs should be able to mimic/block the effects of the cGMP-elevating agents in this intact cell system,
- introduction of active cGMP-PK holoenzyme or fragments by microinjection or transfection methods should be able to mimic the effects of cGMP-elevating agents whereas inactive cGMP-PK mutants should not,
- effects of cGMP-elevating agents should be absent in cGMP-PK-deficient systems, and
- ultimately, the cGMP-PK substrates involved need to be identified.

Unfortunately, there are only very few examples for which most of these criteria have been met. Often, the mediating role of cGMP-PK for a given effect/function is implied or excluded by the use of cGMP-PK activators and/or inhibitors which alone is clearly insufficient to establish or rule out functional roles of cGMP-PKs. Although reasonably well-characterized cGMP-PK activators are available, there is a major lack of reagents which specifically and potently inhibit cGMP-PK activity in intact cells (Smolenski et al. 1998). Also, very often no rigorous examination is made of whether the cGMP-PKs are even expressed in the preparation being investigated, although this is now possible by conventional biochemical and molecular biology techniques. Even when endogenously expressed under *in vivo*-conditions, cGMP-PKs tend to be down-regulated or lost during cell preparation and prolonged cell culturing (Lincoln et al. 1995a; Draijer et al. 1995). Furthermore, long-term treatment with NO or cyclic nucleotides was re-

Table 1. Cellular localization, suggested functions and substrates of cGMP-PK I / cGMP-PK II in mammalian cells

cGMP-PK I

Major Localization	Suggested Functions	Substrates
Brain / Neurons (primarily in cerebellar Purkinje cells, but also spiny striatal neurons, dorsal root ganglion, olfactory bulb and epithelium)	↓ Ca^{2+}_i (?)	G-substrate, DARPP-32 [1], Mena
Smooth muscle cells and related cells (i.e., pericytes, mesangial cells, contractile fibroblasts)	↓ Ca^{2+}_i, contractility (↓ motility, proliferation ?) regulation of gene expression	IP_3R [1], PLB [1], VASP [1], (K_{Ca} channel)
Platelets	↓ adhesion, aggregation, secretion, Ca^{2+}_i, ↓ cytoskeletal reorganization	VASP [1], IP_3R, thromboxane receptor
Cardiac myocytes	↓ contractility ↓ I_{Ca} of L-type Ca^{2+} channel ↓ g_j gap junction conductance	L-type Ca^{2+} channel, connexin 43 [1]
Endothelial cells (not HUVECS)	↓ Ca^{2+}_i, paracellular permeability	VASP [1]
Others (granulocytes, lymphocytes, cell lines)	regulation of gene expression	VASP, vimentin [1], CREB

cGMP-PK II

Major Localization	Suggested Functions	Substrates
Intestinal mucosa	↑ Cl^- secretion (via CFTR)	p25, CFTR [1]
Kidney (JG cells, PT, ATL)	↑ natriuresis, diuresis ↓ renin secretion	(?)
Brain (most brain regions)	↓ vasopressin, neurotransmitter release (?)	(?)
Bone	↑ growth	(?)

[1] cGMP-PK substrates identified also with intact cell preparations. Note that a direct correlation between the substrates and functions listed is not implied, although some examples in which this is the case are discussed in the text.

↑ increase

↓ decrease

Additional abbreviations used in this table: ATL ascending thin limb; Ca^{2+}_i intracellular cytosolic calcium concentration. CREB cAMP response element binding protein; DARPP-32 dopamine and cAMP-regulated phosphoprotein of Mr 32 kDa; IP_3R inositol 1,4,5-trisphosphate receptor; JG juxtaglomerular cells; PLB phospholamban; PT proximal tubules. Modified from Lohmann et al. (1997)

cently shown to down-regulate the expression of cGMP-PK I in smooth muscle cells (Soff et al. 1997).

The role of cGMP-PKs in mediating some important physiological/pharmacological effects of cGMP-elevating agents according to the criteria mentioned above has been established in human platelets (cGMP-PK type I) and intestinal epithelial cells (cGMP-PK type II). In human platelets which contain very high levels (about 4 μM) of cGMP-PK I (Eigenthaler et al. 1992), NO and other cGMP-elevating agents stimulate cGMP-PK activity [as measured by VASP (vasodilator-stimulated phosphoprotein) phosphorylation] and inhibit agonist-evoked calcium transients and subsequent platelet activation. These NO effects are mimicked by cGMP-PK activators, blocked by cGMP-PK inhibitors, and blunted in cGMP-PK-deficient cells (Geiger et al. 1992 and 1994; Butt et al. 1993; Eigenthaler et al. 1993; Butt et al. 1994a,b; Lohmann et al. 1997). However, the precise molecular mechanism of cGMP-PK-inhibition of calcium transients remains to be elucidated. Considerable evidence suggests that cGMP/cGMP-PK-mediated inhibition of agonist-evoked calcium transients, an effect also observed in other cell types (Ruth et al. 1993; Lincoln et al. 1995a; Draijer et al. 1995), is due to the inhibition of phospholipase C and inositol-1,4,5-trisphosphate generation (Waldmann and Walter 1989; Geiger et al. 1992 and 1994; Ruth et al. 1993). The hypothesis that cGMP-PK I causes inhibition of phospholipase C via G-protein phosphorylation (Ruth et al. 1993; Pfeifer et al. 1995) has not been followed up and has not been demonstrated for intact human platelets or other relevant cell types. Recently, cGMP/cGMP-PK-mediated inhibition of human platelets was suggested to involve phosphorylation of the thromboxane receptor and inhibition of thromboxane-specific GTPase (Wang et al. 1998). Unfortunately, cGMP/cGMP-PK inhibition of thromboxane-specific GTPase was demonstrated with platelet membranes, whereas thromboxane receptor phosphorylation in intact cells was only observed using an unusually high concentration (10 mM) of 8-Br-cGMP, and HEL cells in which the presence of cGMP-PK was not demonstrated. Clearly, further work is required to conclusively establish that phosphorylation of platelet agonist receptors by cGMP-PK is functionally relevant.

For intestinal epithelial cells, all of the above mentioned criteria have been met to support the conclusion that membrane-bound cGMP-PK II mediates the effects of cGMP-elevating heat-stable enterotoxins and guanylin on cystic fibrosis transmembrane conductance (CFTR) activation and chloride secretion, which involves cGMP-PK II-mediated CFTR phosphorylation (French et al. 1995; Vaandrager and DeJonge 1996; Pfeifer et al. 1996; Vaandrager et al. 1997 and 1998). This subject has been recently extensively reviewed (Lohmann et al. 1997). Experiments demonstrated that cGMP-PK

II, but neither cGMP-PK I nor cAMP-dependent protein kinase (cAMP-PK), could phosphorylate and activate CFTR in intestinal cells in response to cGMP-elevating agents. Examination of cGMP-PK II-deficient mice confirmed these results (Pfeifer et al. 1996).

There are other examples in which the physiological/pharmacological effects of cGMP were demonstrated to be mediated by cGMP-PK according to most of the criteria described above. Recent experiments strongly suggest that cGMP-PK II inhibits cAMP-stimulated renin secretion in renal juxtaglomerular cells (Gambaryan et al. 1996 and 1998). Other examples will be discussed in the following sections, particularly in sections 4.1 and 4.6 on gene expression.

3 Regulation of Cell Adhesion and Migration by cGMP and cGMP-PKs: an Interplay Between Membrane Receptors and Intracellular Signaling Proteins

Adhesion, migration and proliferation are the basis for many physiological and pathological processes such as embryonal development, wound healing, immune responses, tumor invasion, metastasis formation, and cardiovascular diseases including arteriosclerosis. These cellular functions are regulated by a complex interplay between extracellular adhesive ligands, plasma membrane adhesion receptors, cytoskeletal signaling proteins, and intracellular signaling molecules. Figure 1 shows a scheme of some intracellular signaling cascades and proteins proposed to interact in the regulation of adhesion receptors. cGMP-PKs are increasingly implicated in the control of such signaling mechanisms, although most of the target proteins by which they influence cell motility are still obscure. One major substrate of both cGMP-PK and cAMP-PK is the focal-adhesion associated protein VASP which interacts with various other cytoskeletal proteins and may coordinate interactions between different signaling pathways (Fig. 1). Functional evidence indicates that VASP is a crucial factor involved in the enhancement of actin filament formation (Reinhard et al. 1998). Another possible link between cGMP-PKs and adhesion receptor regulation is the myosin light chain kinase (MLCK) and/or Rho-kinase signaling system. Contractility in muscle and non-muscle cells is controlled by myosin light chain (MLC) phosphorylation resulting from activation of MLCK, partly in response to the Rho-GTP stimulated Rho-kinase (Fig. 1). Dephosphorylation of MLC is catalysed by myosin phosphatase which binds to Rho and is inhibited when phosphorylated by

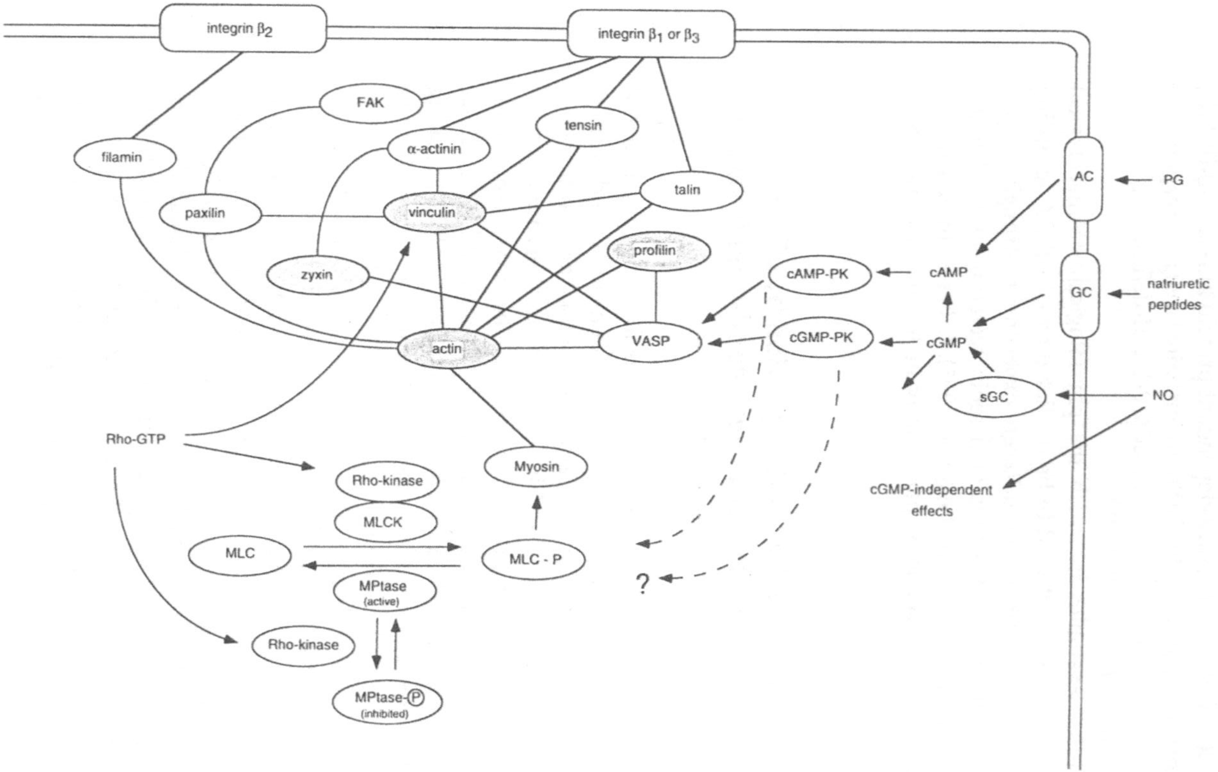

Fig. 1. Protein-protein interactions between integrin adhesion receptors, cytoskeletal proteins, and signaling proteins, and their regulation by cyclic nucleotide/cGMP-PK dependent signal transduction cascades. Lines indicate interactions between proteins, usually based on in vitro assays (reviewed in Yamada and Geiger 1997), gray ovals indicate proteins interacting with VASP, a common substrate of cGMP-PK and cAMP-PK. VASP could function as a mediator of cyclic nucleotide-evoked inhibition of integrin function. Contractility and interaction of signaling molecules is also regulated by MLC phosphorylation stimulated either by MLCK or Rho-kinase, which is activated by Rho-GTP. In response to MLC phosphorylation, myosin assembles into bipolar filaments and generates tension on actin. MLC is dephosphorylated by myosin phosphatase (MPtase), which is inhibited when phosphorylated by Rho-kinase (reviewed in Burridge et al. 1997). Rho further induces a $PI(4,5)P_2$ -mediated conformational change in vinculin allowing it to bind talin and actin. Regulation of MLC phosphorylation/dephosphorylation is a potiential action of cGMP-PK (discussed in the text). Abbreviations used only in this figure: AC adenylyl cyclase; sGC soluble guanylyl cyclase; MPtase myosin phosphatase; PG prostaglandins

Rho-kinase (reviewed in Burridge et al. 1997). Both cGMP- and cAMP-elevating vasodilators have been shown to inhibit agonist-induced MLC phosphorylation in vascular smooth muscle cells (SMCs), fibroblasts and platelets (Walter 1989). In human platelets, this effect appears to be secondary to inhibition of agonist-stimulated PLC or some other site upstream of the activation of the two divergent pathways of MLCK and protein kinase C (Waldmann and Walter 1989; Walter 1989; Geiger et al. 1992). There is very little experimental evidence supporting the direct phosphorylation and inhibition of MLCK by cGMP-PK in intact cells which would result in decreased MLC phosphorylation. Possibly, cGMP and cGMP-PK activate a myosin light chain phosphatase (Lee et al. 1997; Wu et al. 1998). A cGMP/cGMP-PK regulated phosphatase has also been implicated as mediator of cGMP-evoked stimulation of calcium-activated potassium channels (Zhou et al. 1996; Sansom et al. 1997). The cGMP/cGMP-PK regulation of the MLC/MLCK system may have important functional consequences for the adhesive properties of cells since MLC phosphorylation leads to assembly of myosin into bipolar filaments that increase actin tension. This increased tension appears to cluster integrin receptors, activate focal adhesion kinase (FAK) and recruit various signaling proteins into focal adhesions (Burridge et al. 1997). Interestingly, myosin has recently been identified as a ubiquitous cGMP-PK binding protein (Vo et al. 1998). In addition, Rho induces a phosphatidylinositol (4,5)-bisphosphate ($IP(4,5)P_2$)-mediated conformational change in vinculin that enables it to bind to talin and actin, which increases the link between integrins and actin filaments (Fig. 1) (reviewed in Burridge et al. 1997).

Here, the role of cGMP-PK in regulation of adhesive and migratory cell processes will be considered with particular emphasis on cardiovascular cells. Other effects of cGMP-PK on cell growth cannot be discussed in detail in this review. Increased proliferation and motility of cardiovascular cells are key events in the pathogenensis of atherosclerosis and coronary heart disease. Cyclic nucleotide-elevating nitrovasodilators have wide use in therapy of cardiovascular diseases, therefore the molecular basis of their short-term, and especially long-term effects are of interest and will be discussed in several sections dealing with cells of the vessel wall [e.g. vascular SMCs, endothelial cells (ECs)] and blood cells (e.g. platelets, leukocytes). Each cell type contains an individual pattern of cyclic nucleotide-regulated protein kinases and other cGMP targets. Since NO can regulate cell functions in cGMP-independent ways (Peng et al. 1995; Ignarro and Murad 1995), we will focus primarily on changes in cell motility and proliferation caused by cGMP-elevating natriuretic peptides and/or membrane-permeable cGMP analogs. NO effects will be mentioned when there is additional evidence that at least some of the effects of NO are mediated by cGMP.

3.1 Adhesion and Migration of Vascular Endothelial Cells

Adhesion, migration and proliferation of ECs are involved in both constructive angiogenesis and neovascularization but also in the initiation and progression of destructive processes underlying various vascular diseases. ECs possess a large number of GC-coupled ANP receptors (Itoh et al. 1992). Components of the cGMP signaling system including cGMP-PK and its substrate VASP are present in most but not all ECs (Draijer et al. 1995; Markert et al. 1996). cGMP-PK type I was demonstrated in freshly isolated ECs from human aorta and iliac arteries as well as in cultured ECs from human aorta, iliac vein, and foreskin microvessels (Draijer et al. 1995). However, both expression of this enzyme and cGMP-dependent VASP phosphorylation decrease during serial passage of ECs. Also, in small capillaries of the human heart, high expression levels and colocalization of cGMP-PK and VASP were observed (Markert et al. 1996). However, the widely used human umbilical vein ECs (HUVECs) do not contain detectable levels of cGMP-PK type I, and cGMP analogs also do not induce VASP phosphorylation in these cells (Draijer et al. 1995; Smolenski A et al. unpublished). cGMP analogs are able to inhibit thrombin-induced increase in EC permeablility only in ECs containing cGMP-PK I, suggesting that this effect is mediated by cGMP-PK (Draijer et al. 1995). Therefore, effects of cGMP-analogs observed in HUVEC cultures most likely result from other signaling pathways or even from contaminating cell populations such as fibroblasts.

Angiogenesis is a complex process of growing new capillary blood vessels from preexisting capillaries and postcapillary venules. In this process, adhesion receptors play a major role in EC proliferation, migration, differentiation, and organization into patent capillary networks (Bischoff 1997). Recently, it has been elegantly demonstrated that integrin adhesion receptor subunits α_v, β_1, β_3 and β_5 play an important role in angiogenesis. The knockout of the integrin β_1 chain in mice produced a phenotype with disrupted blood vessel formation (Bloch et al. 1997). Integrin $\alpha_v\beta_3$ appears to be a crucial component in the formation of new tumor blood vessels, since antibodies against this receptor efficiently block tumor-induced angiogenesis in several *in vivo* models (Brooks et al. 1994; Friedlander et al. 1995).

Although cGMP-PK clearly is involved in integrin regulation in other cell types (see subsequent paragraph 3.4), few studies in ECs have addressed the question of whether NO/cGMP are signals for cell motility and angiogenesis.

Migration (as well as proliferation) of cultured ECs can be stimulated by various factors, e.g. thrombin or serum. A variety of evidence indicate that cGMP-regulated signaling pathways oppose these effects. Serum-stimulated migration of rat vascular ECs in Boyden chambers was inhibited by cardiac natriuretic peptides (rat ANP1-28 and rat BNP-45) as well as by 8-Br-cGMP (Ikeda et al. 1995). In cultured bovine aortic ECs, ANP and 8-Br-cGMP attenuated serum-stimulated DNA synthesis and reduced the increase in cell number. Furthermore, the proliferative action of basic fibroblast growth factor on ECs was attenuated by 8-Br-cGMP (Itoh et al. 1992). NO donors [S-nitrosopenicillamine (SNAP), sodium nitroprussied (SNP)] as well as 8-Br-cGMP inhibited serum-induced migration of cultured HUVECs (Lau and Ma 1996), but as discussed above, this effect most likely does not involve cGMP-PK.

In a complex *in vitro* angiogenesis model that involves growing rat microvascular ECs in three-dimensional gels and allowing them to form tubular systems in the presense of tumor growth factor beta 1 (TGFβ1), L-NAME [an inhibitor of nitric oxide synthase (NOS)] reduced tube formation and accumulation of nitrite (Papapetropoulos et al. 1997a). Addition of SNP or 8-Br-cGMP to the gel restored capillary tube formation. In contrast, in two-dimensional cultures, inhibition of NOS by L-NAME did not influence EC proliferation or antagonize the ability of TGFβ1 to suppress EC proliferation (Papapetropoulos et al. 1997a). Vascular endothelial growth factor (VEGF) is a specific growth factor for ECs with strong angiogenic activity *in vitro* and *in vivo*. Exposure to VEGF stimulated NO production and release, and increased intracellular cGMP levels in coronary venular ECs and other EC types (Ziche et al. 1997, Papapetropoulos et al. 1997b). Experiments with NOS and GC inhibitors demonstrated that NO and cGMP mediate some of the effects of VEGF on EC proliferation, migration, adhesion and angiogenesis (Ziche et al. 1997, Parenti et al. 1998). Both short-term and long-term exposure to VEGF stimulated growth and promoted NO-dependent formation of network-like structures in three dimensional collagen gels (Papapetropoulos et al. 1997b). Systemic administration of a NOS inhibitor (L-NAME) in a rabbit corneal implant model blocked VEGF, but not basic fibroblast growth factor-induced angiogenesis (Ziche 1997). In addition to serving as down-stream mediators of VEGF functions, NO and cGMP may also regulate VEGF mRNA expression and stability in human glioblastoma and hepatoma cells (Chin et al. 1996).

Pathophysiological conditions *in vitro* and *in vivo*, such as hypoxia or thrombin stimulation of ECs, increase platelet adhesion to ECs. This interaction between platelets and ECs is mediated by adhesion receptors, one of which is the adhesion molecule P-selectin, found in secretory granules of both platelets and ECs. Interestingly, P-selectin expression on the EC surface appears to be diminished by endogenous NO and also by 8-Br-cGMP (Davenpeck et al. 1994). This may constitute another important anti-adhesive mechanism, in addition to NO inhibition of platelet function. Other adhesion molecules on ECs are also down-regulated by NO or cGMP. Tumor necrosis factor alpha (TNFα)-induced vascular cell adhesion molecule-1 (VCAM-1) expression on ECs is inhibited by NO-donors, 8-Br-cGMP and flow-induced endothelial NO release (Tsao et al. 1996). NO and nitrovasodilators also inhibited the IL-1-induced increase in intercellular adhesion molecule-1 (ICAM-1) and VCAM-1, an effect blocked by hemoglobin inhibition of NO (De Caterina et al. 1995; Takahashi et al. 1996). However, 8-Br-cGMP did not have the same effect as the nitrovasodilator, indicating that NO donor effects may be mediated through cGMP/cGMP-PK-independent mechanisms such as NO interaction with heme proteins, non-heme iron, or free thiol residues on target signaling proteins, enzymes or ion channels (Schmidt and Walter 1994; De Caterina et al. 1995; Takahashi et al. 1996).

Thus, the available data suggests that the NO/cGMP system has an inhibitory influence on EC adhesion, migration and proliferation, but is involved in the regulation of angiogenesis by VEGF. However, the data in HUVECs suggest some functions of NO/cGMP independent of cGMP-PK. Despite the potential role of cGMP-PK in integrin regulation and the crucial role of integrins in cell migration during angiogenesis, the link between these two signaling systems has not been reported.

3.2 Adhesion and Migration of Vascular Smooth Muscle Cells

It is well established that cGMP-elevating agents promote smooth muscle relaxation (Walter 1989; Lincoln et al. 1995a). Recent experiments using cGMP-PK type I-deficient mice demonstrate that this kinase mediates the inhibitory effect of cGMP-elevating agents on smooth muscle contraction (Pfeifer et al. 1998). In pathology, vascular SMCs play an important role in the later phase of coronary heart disease. Both vascular SMC proliferation and migration contribute to atherosclerotic lesion formation, and migration of medial smooth muscle cells into the intima is considered to be an important process of intimal thickening (Schwartz 1997). SMCs contain all essential components of the cGMP signaling system. However, in culture, SMCs loose their cGMP-PK after a few passages and therefore constitute an ideal model for studying effects occuring in the presense and absence of cGMP-PK.

In femoral artery and human heart, medial SMCs, small capillaries, and ECs were strongly positive for cGMP-PK and its substrate VASP by immunofluorescence (Markert et al. 1996). Immunohistochemistry and immunoblotting experiments on denuded rat carotid arteries demonstrated a homogenous distribution of cGMP-PK I throughout the neointima, while VASP was more concentrated in vascular SMCs lining the artery lumen (Mönks et al. 1998). This pattern might reflect enhanced growth factor stimulation of luminal vascular SMCs in the neointima, and also suggested the involvement of both cGMP-PK I and its substrate VASP in vessel wall regulation and inhibition of restenosis formation.

Several *in vitro* studies concerning the role of the cGMP/cGMP-PK system in vascular SMCs have been performed. In a series of elegant experiments, cGMP-PK function was investigated by restoring cGMP-PK expression to vascular SMCs that had become cGMP-PK deficient during prolonged culture (Boerth et al. 1997, Dey et al. 1998). Stable transfection of cGMP-PK Iα or Iβ into these subcultured, kinase-deficient vascular SMCs resulted in cGMP-PK expression levels comparable to those observed in freshly isolated vascular SMCs and transformed these de-differentiated, cGMP-PK-deficient vascular SMCs from a 'synthetic' to a 'contractile' morphology. Expression of cGMP-PK resulted in increased production of contractile phenotype marker proteins such as smooth muscle myosin heavy chain-2, calponin and alpha-actin, and inhibited the synthesis and secretion of two extracellular matrix proteins, osteopontin and thrombospondin, which are markers for the synthetic phenotype and involved in the formation of neointima. Expression of cGMP-PK restored the capacity of cGMP and cAMP analogs to inhibit platelet-derived growth factor (PDGF)-induced cell migration, although it had little effect on PDGF-induced cell proliferation. Furthermore, comparison of low and high passage vascular SMCs and transfection of cGMP-PK Iα into cells that had lost their cGMP-PK activity also demonstrated that this kinase is necessary but not sufficient to trigger focal adhesion disassembly in intact rat aortic SMCs (Murphy-Ullrich et al. 1996).

In a Boyden chamber assay, stimulated migration of human coronary artery SMCs was inhibited by natriuretic peptides (ANP, BNP, CNP-22), cGMP analogs (8-Br-cGMP) and NO donors (Kohno et al. 1997). In the same study there was no detectable suppression of cell adhesion by these substances. All natriuretic peptides increased intracellular cGMP levels and acted, at least in part, through a cGMP-dependent mechanism. ANP, as well as nitrovasodilators and 8-Br-cGMP, also exerted an antigrowth effect on vascular SMCs via GC-coupled mechanisms (Garg and Hassid 1989, Itoh et al. 1990).

NO-donors and 8-Br-cGMP inhibited angiotensin II-induced migration of rat aortic SMC in Boyden chambers (Dubey et al. 1995). In the presense of L-arginine, but not D-arginine, the iNOS inducer, Il-1ß, inhibited angiotensin II-induced vascular SMC migration, and this was prevented by a NOS inhibitor. The effects of NO donors in this system were partially inhibited by LY83583 (inhibitor of GC) or KT5823 (inhibitor of cGMP-PK), however, the specificity and potency of these latter substances is unproven in intact cells (discussed in Smolenski et al. 1998). Furthermore, in this study 8-Br-cAMP also displayed antimigratory effects, but NO donor effects were not altered by inhibitors of either adenylyl cyclase or cAMP-PK. In wounding assays, which serve as an additional approach to studying SMC migration, a confluent monolayer of cells is disrupted by scratching away the cells, and the migration of the remaining cells into the cell-free area is observed. As in the Boyden chamber analysis, nitrovasodilators and 8-Br-cGMP reversibly inhibited the number of migrating vascular SMCs, and the distance they traveled (Sarkar et al. 1996).

A variety of animal models demonstrated that augmentation of NO production *in vivo* decreases intimal lesions in arterial injury, whereas inhibition of NOS has the opposite effect. In a recent study, rat arterial SMCs were infected with a retrovirus expressing ecNOS, then seeded back onto the luminal surface of balloon-injured rat carotid arteries. The NOS overexpressing cells displayed increased VASP phosphorylation, markedly inhibited neointima formation and induced dilatation of the vessel compared to vessels seeded with control cells (Chen et al. 1998). Furthermore, *in vitro* cell proliferation and DNA synthesis were inhibited in the ecNOS-expressing vascular SMCs. The antiproliferative effects of NO in vascular SMCs may be partially mediated by cGMP and cGMP-PK (Yu et al. 1997). However, amplification rather than inhibition of certain growth factor responses by cGMP has been reported (Dhaunsi and Hassid 1996).

In spontaneously hypertensive rats (SHR), evidence suggests that intrinsic abnormalities of the NO/cGMP pathway may exist, such as low cardiac cGMP-PK (Kuo et al. 1976, Ecker et al. 1989) and abnormal L-arginine metabolism (Hasegawa et al. 1992). In SHRs, chronic L-arginine administration attenuated both cardiac hypertrophy (decreased the heart/body weight ratio), and the expression of skeletal alpha-actin mRNA in cardiac myocytes (Matsuoka et al. 1996), however, the contribution of cGMP-PK to these effects has not been clarified.

In conclusion, cGMP-PK expression is involved in inhibition of migration and in sustaining a contractile-like phenotype in cultured vascular SMCs. cGMP-PK and its substrate VASP are major targets of vasodilatory agents and are likely to regulate cell motility during restenosis and cardiovascular diseases. However, definition of the precise molecular links between cGMP-PK and effects on vascular SMC motility, adhesion and proliferation require further investigation.

3.3
Adhesion and Migration of Leukocytes

Leukocytes/neutrophils play an important role in wound healing, immune responses, and tumor defense. However, activated leukocytes also release oxygen radicals that can enhance tissue damage in cardiovascular diseases. The initial step in the migration of neutrophils to the extravascular space is adhesion to the endothelium. The interaction of leukocytes and ECs involves several adhesion receptors. Initial contact, via selectins, is characterized by tethering and rolling of leukocytes (McEver and Cummings 1997). After this, stronger contact is mediated by other adhesion receptors such as β_2 integrins.

Leukocytes contain minor amounts of cGMP-PK in comparison to platelets and vascular SMCs. In our own experiments, we were unable to detect cGMP-PK by the standard Western blot technique, nor by the very sensitive method of cGMP-mediated VASP phosphorylation (Klippel et al. to be published). However, cGMP-PK was detected in leukocytes by immunoprecipitation, and immunofluorescence studies showed diffuse staining in the cytoplasm, at the microtubule organizing center and in the euchromatin of the nucleus (Pryzwansky et al. 1990 and 1995b). In fMLP- or Ca^{2+} ionophore-stimulated neutrophils, cGMP-PK transiently redistributed to specific targeting proteins or structures of the cytoskeleton, perhaps vimentin (Pryzwansky et al. 1990 and 1995b). Adhering leukocytes demonstrated a transient increase in cGMP levels that was coincident with co-localization of cGMP-PK and vimentin, as well as increased phosphorylation of vimentin (Wyatt et al. 1991).

A role for cGMP in neutrophil adhesion, migration and degranulation has been proposed, but the intracellular mechanisms of cGMP action are poorly understood. Short term adhesion of neutrophils to collagen I can be reduced by the NO-donor SNAP and 8-Br-cGMP (Sundqvist et al. 1994). After prolonged incubation there was no difference between SNAP and untreated cells suggesting that the effect was mediated by NO which declines as the short-lived SNAP deteriorates.

Several studies have been performed on the migration of electropermeabilized neutrophils using the Boyden chamber. Various activators of neutrophil migration increased intracellular cGMP levels, and the effects on migration could be attenuated or blocked by inhibitors of GC or cGMP-PKs, suggesting a stimulatory effect of cGMP on neutrophil migration. However, it is not clear whether electropermeabilized neutrophils have the same response to external stimuli as non-electropermeabilized cells. Maximal stimulation of neutrophil migration appeared only at a very specific concentration of cGMP, whereas higher concentrations were less effective, or even inhibitory (Elferink and de Koster 1993, Elferink and vanUffelen 1996).

In contrast, other studies reported an inhibitory effect of luminally-released endothelial NO on leukocyte adhesion and the expression of adhesion molecules (Bassenge 1994). Inhibition of NO-synthesis by various NOS inhibitors increased leukocyte oxidant release and stimulated rolling and adhesion of neutrophils to ECs via mast cell-/oxidant-dependent mechanisms (Davenpeck et al. 1994; Niu et al. 1994 and 1996). NO donors and 8-Br-cGMP attenuated some of the NOS inhibitor effects. Adhesion of neutrophils could further be blocked by anti-CD18 or anti-ICAM antibodies. Adhesion between leukocytes and platelets is also influenced by NO/cGMP-pathways. NOS inhibitors (e.g. L-NAME) can induce aggregation between platelets and leukocytes, an effect that is attenuated by 8-Br-cGMP, as well as by antibodies to P-selectin (Kurose et al. 1993).

In conclusion, the effects of cGMP on leukocyte adhesion and migration may very much depend on the concentration of cGMP, the cell system used for investigation, and also the cell type or extracellular matrix the leukocytes adhere to. More studies will be required, and the use of leukocytes from cGMP-PK I-deficient mice might help to elucidate the role of the cGMP/cGMP-PK system in these cells.

3.4 Inhibition of Platelet Function by cGMP-Dependent Signaling Cascades

One of the best studied cardiovascular cell types with respect to cGMP-PK function are human platelets (see also section 2). Platelet adhesion and activation are normally prevented by anti-adhesive mechanisms which may be impaired under pathophysiological conditions. Platelet aggregation is prevented by both cGMP- and cAMP-elevating vasodilators such as prostacyclin and EDRF/NO which are constantly released from vascular ECs (Nolte et al. 1991; Schmidt and Walter 1994). Platelets contain high levels of both cGMP- and cAMP-dependent protein kinases (Eigenthaler et al. 1992), and the molecular mechanisms of their action in platelets most likely involves several steps in the platelet activation cascade such as phospholipase C, MLCK, calcium mobilization and changes in the cytoskeleton (Waldmann and Walter 1989; Geiger et al. 1992,1994; Horstrup et al. 1994; Lohmann et al. 1997).

An important mediator of effects of cyclic nucleotide dependent protein kinases in platelets is likely to be the protein VASP, a substrate with three distinct but differentially preferred phosphorylation sites for cGMP-PK and cAMP-PK (Butt et al. 1994b). Cyclic nucleotide-dependent phosphorylation of VASP correlates very well with inhibition of integrin $\alpha_{IIb}\beta_3$ (fibrinogen receptor) activation and binding of soluble fibrinogen to this receptor, the key event in platelet aggregation (Horstrup et al. 1994; Eigenthaler and Shattil, 1996; Yamada and Geiger 1997). Unlike inhibition of platelet aggregation by both cAMP and cGMP, platelet adhesion to extracellular matrix proteins or endothelial cells can be completely inhibited only by cGMP-elevating vasodilators, cGMP analogs or cGMP-PDE inhibitors, whereas cAMP-elevating substances or cAMP analogs only inhibit partially (Radomski et al. 1987; Venturini et al. 1992). These *in vitro* results were confirmed using an *in vivo* rat carotid artery model system (Vemulapalli et al. 1996). Newly developed direct activators/modulators of GCs like YC-1 may therefore be powerful inhibitors of both initial and long-term platelet adhesion to collagen, and their inhibitory effect may even be potentiated by selective inhibitors of cGMP-specific phosphodiesterases such as dipyridamole (Wu et al. 1997). Combination of these drugs which have distinct effects on cGMP synthesis and degradation could become an important pharmacological approach for inhibiting platelet adhesion *in vivo*.

Platelets from certain patients with chronic myelocytic leukemia (CML) show very little if any expression of cGMP-PK, and have been an interesting model system for studying cGMP-regulated signaling cascades. In cGMP-PK-deficient platelets, SNP-stimulated protein phosphorylation and cGMP analog inhibiton of agonist-evoked calcium response were severely impaired, whereas the cAMP/cAMP-PK response was intact (Eigenthaler et al. 1993). Since availability of such platelets is very limited, platelets from cGMP-PK I-deficient mice (Pfeifer et al. 1998) will be very useful for further biochemical, physiological and pharmacological analysis of the relevance of cGMP-PK for platelet function *in vivo* .

4
cGMP, cGMP-PK, and Gene Expression

Compared to our knowledge about the regulation of gene expression by the cAMP/cAMP-PK signal transduction pathway, our understanding of the regulation of gene expression by cGMP/cGMP-PK is only now emerging. Since NO can also regulate gene expression in cGMP-independent ways (Peng et al. 1995; Ignarro and Murad 1995), we will concentrate on changes in gene expression caused by cGMP-elevating natriuretic peptides and/or membrane-permeable cGMP analogs. Tables 2 and 3 summarize some of the data published in mammalian cells, although regulation of gene expression by cGMP has also been documented in lower organisms and plants (Bowler et al. 1994). Both activation and inhibition of gene expression by cGMP has been observed. In many cases, cGMP regulates gene expression at the transcriptional level, but post-transcriptional regulation by cGMP has also been reported. First, we will discuss those genes for which there is clear demonstration that the effect of cGMP on gene expression is mediated by cGMP-PK, because it was dependent on expression of transfected cGMP-PK in cGMP-PK-deficient cells (summarized in Table 2). Secondly, we will discuss examples in which the cGMP effect on gene expression was abrogated by a cGMP-PK inhibitor suggesting involvement of the kinase, as well as other examples in which the data do not allow any conclusion as to whether the effect of cGMP was mediated by cGMP-PK (Table 3). Since many of the genes listed in Tables 2 and 3 are also regulated by cAMP-PK, the potential for cGMP cross-activation of cAMP-PK must be considered in each case. Thirdly, we will discuss some of the mechanisms involved in transcriptional regulation by cGMP-PK, including nuclear translocation of the kinase and the *cis*-acting regulatory elements and *trans*-acting transcription factors which are targets of the kinase. Finally, we will briefly mention some of the physiological functions of cGMP-PK which are likely to involve regulation of gene expression.

4.1
Regulation of Gene Expression by cGMP-PK I

The *fos* and *junB* gene products are part of the AP-1 (Fos/Jun) transcription factor complex which plays an important role in regulating cell growth and differentiation. Transcription of *c-fos* and *junB* is induced rapidly by a variety of stimuli including activation of cAMP-PK, protein kinase C, or Ca^{2+}/calmodulin-dependent protein kinase (Angel and Karin 1991). In many different cell lines, cGMP-elevating agents and membrane-permeable cGMP analogs transiently increase c-*fos* and *junB* mRNA levels, with *c-fos* and *junB* expression peaking at 30 min and 1 h, respectively (Pilz et al. 1995; Haby et al. 1994; Thiriet et al. 1997; Belsham et al. 1996; Uberall et al. 1994). In PC-12 cells, one group reported that cGMP only amplified Ca^{2+}-induced increases in c-*fos* mRNA (Peunova and Enikolopov 1993), whereas others reported induction of *c-fos* by cGMP analogs alone (Haby et al. 1994). The reason for this discrepancy is unclear, although different growth conditions of the cells might influence *c-fos* promoter activity and responsiveness to stimuli.

Table 2. Regulation of mRNA / Protein Expression by cGMP-PK

Gene	Cell Type / Tissue	Observed Effects of cGMP	References
c-fos	BHK cells, transfected with cGMP-PK A10 smooth muscle cells	Increased *c-fos* promotor activity [2,3]	Gudi1996 unpublished [6]
	PC12 phaeochromocytoma cells [1] NIH 3T3 fibroblasts [1]	Increased *c-fos* promotor activity (co-stimulation with Ca^{2+} ionophore or neopterin)	Peunova 1993 Überall 1994
c-fos, junB	BHK cells transfected with cGMP-PK REF 52 fibroblasts Rat thyroid cells PC12 phaeochromocytoma cells [1] C6 glioma cells GT1 hypothalamic cells	Increased mRNA [2,4]	unpublished [6] Pilz 1995 Pilz 1995 Haby 1994 Thiriet 1997 Belsham 1996
Myosin heavy chain-2 Calponin α-Actin	Passaged rat aortic VSMC, transfected with cGMP-PK	Increased protein (mRNA not examined) [2]	Boerth 1997
Thrombospondin Osteopontin	Passaged rat aortic VSMC, transfected with cGMP-PK	Decreased protein [2], no change in mRNA	Dey 1998
Tumor necrosis factor (TNF)-α	Rat peritoneal macrophages BHK cells transfected with cGMP-PK	Increased mRNA [5] Increased TNF-α promotor activity [2]	Gong 1990 unpublished [6]
MAP-kinase phosphatase 1	Rat glomerular mesangial cells BHK cells transfected with cGMP-PK	Increased mRNA [2]	Sugimoto 1996 Suhasini 1998

[1] cGMP-PK expression was not directly demonstrated in these cells
[2] Effect in cGMP-PK deficient cells is dependent on transfected cGMP-PK I
[3] Effect is not abrogated by cAMP-PK inhibitor
[4] Effect is abrogated by cGMP-PK inhibitor
[5] cAMP has no effect or opposite effect
[6] Gudi T and Pilz RB, unpublished results

Table 3: Regulation of mRNA / Protein Expression by cGMP, Possible Involvement of cGMP-PK

Gene	Cell Type / Tissue	Observed Effects of cGMP	References
Gonadotropin-releasing hormone	GT1 hypothalamic cells	Decreased mRNA and promotor activity (co-stimulation with Ca 2+ ionophore) [4,5]	Belsham 1996
Microtubule-associated protein 1	Hippocampal granule cells	Increased mRNA [4]	Johnston 1994
Type A atrial natriuretic peptide receptor	Embryonic rat aortic VSMC	Decreased mRNA and promotor activity [5]	Cao 1995
Alkaline phosphatase Osteocalcin Type 1 collagen	Murine and rat osteoblasts (primary culture and cell line)	Increased mRNA [5]	Inoue 1995 and 1996
Soluble guanylyl cyclase ($\alpha 1/\beta 1$ subunits)	Primary rat pulmonary artery VSMC	Decreased mRNA stability	Filippov 1997
Asialoglycoprotein Receptor	HepG2 and HuH-7 human hepatoma cells [1]	Increased protein (increased mRNA translation mediated by RNA binding protein) [5]	Stockert 1997
egr-1	PC12 phaeochromocytoma cells [1] C6 glioma cells	Increased mRNA	Thiriet 1997
E2F-1	Primary rat aortic VSMC,	Decreased mRNA (under hypoxic conditions)	Morita 1997
Cyclooxygenase 2	Rat glomerular mesangial cells	Increased mRNA (potentiation of interleukin 1)	Tetsuka 1996
P-Selectin	Rat mesenteric postcapillary venule endothelial cells	Decreased protein	Davenpeck 1994
α-Actin	Rat hepatic stellate cells [1]	Decreased mRNA	Kawada 1996
Leptin	Rat epididymal fat [1]	Increased mRNA [5]	Yoshida 1996
Elastin	Primary fetal bovine fibroblasts [1]	Increased protein [5]	Mecham 1985

See Table 2 for footnotes.

Transcriptional regulation of the *fos* promoter by cGMP has been systematically examined (Peunova and Enikolopov, 1993; Gudi et al. 1996 and 1997). Using cGMP-PK-deficient baby hamster kidney (BHK) cells transfected with a cGMP-PK Iβ expression vector, it was demonstrated that the effect of cGMP on the c-*fos* promoter was mediated by cGMP-PK (Gudi et al. 1996). The intracellular concentration of cGMP-PK produced in transfected BHK cells was comparable to physiological cGMP-PK concentrations found in smooth muscle cells and platelets (Gudi et al. 1996). Reporter gene expression from the *fos* promoter was induced by 8-Br-cGMP only in cGMP-PK-transfected BHK cells but not in cells transfected with "empty" control vector, and was strictly dependent on both the amount of cGMP-PK vector transfected and the 8-Br-cGMP concentration used. Reporter gene expression from several other promoter constructs was not influenced by cGMP-PK or 8-Br-cGMP, indicating that the effect of cGMP-PK required certain sequence elements of the *fos* promoter [(Gudi et al. 1996) and see section 4.3]. Control experiments with Ca^{2+} ionophores and the specific cAMP-PK inhibitor, PKI, demonstrated that cGMP-PK activation of the *fos* promoter was independent of Ca^{2+} and cAMP-PK in BHK cells (Gudi et al. 1996). In PC12 cells, co-stimulation of the *fos* promoter by cGMP and Ca^{2+} was inhibited by a short PKI-derived peptide (Peunova and Enikolopov 1993), however, short PKI-derived peptide sequences may inhibit both cAMP- and cGMP-PK, whereas full-length PKI inhibits specifically only cAMP-PK, not cGMP-PK (Glass et al. 1986; Kemp et al. 1988).

Recently, regulation of gene expression by cGMP-PK was demonstrated in vascular SMCs which rapidly lose their cGMP-PK activity and undergo phenotypic dedifferentiation when cultured *ex vivo* (Boerth et al. 1997). In comparison to cells transfected with control vector, kinase-deficient vascular SMCs stably transfected with cGMP-PK Iα or Iβ showed increased expression of smooth muscle myosin heavy chain-2, calponin and α-actin [mRNA levels were not examined, (Boerth et al. 1997)]. Also, in comparison to cGMP-PK-deficient control-transfected cells, cGMP-PK-transfected vascular SMCs demonstrated dramatically reduced synthesis and excretion of osteopontin and thrombospondin, without a change in their steady-state mRNA levels, suggesting a post-transcriptional mechanism (Dey et al. 1998). Importantly, the changes in gene expression caused by cGMP-PK correlated with morphological changes indicative of a more differentiated, contractile phenotype and with functional changes, i.e. cGMP-mediated inhibition of cell migration [(Boerth et al. 1997) and see section 3.2].

Regulation of the TNFα gene by cyclic nucleotides is complex since low concentrations of cAMP and cGMP increase TNFα mRNA expression in macrophages, whereas high concentrations of cAMP are inhibitory (Gong et al. 1990). Another study found no effect of cGMP analogs on TNFα mRNA expression, but was performed using cultured cells which may have lost cGMP-PK expression (Wang et al. 1997). Furthermore, the latter study demonstrated that NO increased TNFα expression by inhibiting adenylate cyclase and decreasing intracellular cAMP levels. In cGMP-PK Iβ transfected BHK cells, we found cGMP-dependent induction of the TNFα promoter which was dependent on the presence of the kinase (Gudi T and Pilz RB, unpublished results).

Mitogen-activated protein kinase phosphatase (MKP-1) is a dual specificity phosphatase which attenuates growth factor-induced mitogen-activated protein kinase activity. Induction of MKP-1 mRNA expression by cGMP-elevating agents and cGMP analogs was demonstrated in glomerular mesangial cells and correlated with cGMP-induced growth inhibition (Sugimoto et al. 1996). Treatment of BHK cells with cGMP analogs increased MKP-1 mRNA expression only in cells transfected with cGMP-PK, but not in control-transfected cells, indicating that the effect of cGMP is mediated by cGMP-PK, not by cross-activation of cAMP-PK (Suhasini M et al. 1998).

4.2 Regulation of Gene Expression by cGMP Possibly Involving cGMP-PK

Several studies in cGMP-PK-expressing neuronal cells describe regulation of gene expression by cGMP which is abrogated by cGMP-PK inhibitors, suggesting involvement of the kinase. In hypothalamic cells, repression of the gonadotropin releasing hormone (GnRH) gene by cGMP required elevation of cytosolic calcium, and cGMP-elevating agents or cGMP analogs had no effect in the absence of calcium ionophores (Belsham et al. 1996). The effect of cGMP on GnRH mRNA expression was blocked by the cGMP-PK inhibitor Rp-cGMPS, and activation of cAMP-PK had no effect on GnRH mRNA levels, excluding cross-activation of cAMP-PK as a mechanism (Belsham et al. 1996). In hippocampal granule cells, treatment with cGMP-elevating agents or cGMP analogs increased the level of microtubule-associated protein-2 mRNA, and the putative cGMP-PK inhibitor KT5823 abrogated the effect, however cGMP had no effect on the levels of mRNAs encoding several other cytoskeletal proteins (Johnston and Morris, 1994).

There are several examples of cGMP-regulated gene expression in cGMP-PK-expressing cells in which cAMP analogs were shown to have no effect or the opposite effect of cGMP, but definitive data indicating involvement of cGMP-PK are lacking. These examples include the effects of cGMP-elevating agents and cGMP analogs on natriuretic peptide receptor-A mRNA expression in aortic vascular SMC (Cao et al. 1995), and on alkaline phosphatase, osteocalcin and type 1 collagen mRNA expression in osteoblast-like cells (Inoue et al. 1995 and 1996).

Interesting effects of cGMP on mRNA stability and translation, which may or may not be mediated by cGMP-PK, have been described (Filippov et al. 1997; Stockert and Ren 1997). Decreased GC expression in vascular SMCs treated with 8-Br-cGMP correlated with decreased stability of the mRNAs encoding the α1 and β1 subunits of the enzyme (Filippov et al. 1997). Since cAMP analogs had effects similar to those of cGMP, cross-activation of cAMP-PK cannot be excluded. Increased expression of the asialo-glycoprotein receptor in cGMP-treated hepatoma cells was found to be due to increased mRNA translation mediated by a specific mRNA binding protein. In this system, cAMP and cGMP had opposite effects (Stockert and Ren 1997).

4.3
Transcription Regulatory Elements Targeted by cGMP-PK I

The *fos* promoter contains a complex of several known enhancer elements including a *sis*-inducible element (SIE), a serum-response element (SRE) with an adjacent binding site for the CCAAT/enhancer binding protein C/EBP-β, an AP-1 binding site (FAP), and a cAMP-response element (CRE) (Fisch et al. 1989). Experiments in BHK cells with constructs of *fos* promoter deletions and mutations demonstrated that cGMP-PK effects on the *fos* promoter were mediated by several sequence elements including the SRE, FAP and CRE. Each element could function independently in confering cGMP inducibility to a heterologous promoter, but within the context of the native *fos* promoter, the contribution of each sequence did not appear to be additive (Gudi et al. 1996). Costimulation of the *fos* promoter by NO and Ca^{2+} in PC12 cells was significantly reduced when sequences including the SIE, SRE and FAP elements were removed, and the response was completely lost when the CRE was deleted (Peunova and Enikolopov, 1993). Although transactivation of the *fos* promoter by cGMP-PK in BHK cells was similar in magnitude and required similar cis-acting elements as induction by cAMP-PK, there were significant differences between cGMP- and cAMP-PK activation of single enhancer elements (Gudi et al. 1996).

Comparison of the regulatory sequences of three cGMP-PK-responsive genes, *c-fos*, *junB* and TNFα, demonstrate the presence of similar key regulatory elements. The *junB* promoter contains a CCAAT enhancer element, a CRE-like element, plus a GC-rich region, and additional enhancer elements in the 3'-flanking region of *junB* (including an SRE, CRE and two potential C/EBP-β binding sites) which are important for regulation of *junB* by growth factors, phorbol esters and cAMP (De Groot et al. 1991; Perez-Albuerne et al. 1993). The sequences of the TNFα 5'-flanking region required for cGMP responsiveness contain a canonical CRE, a variant TRE, two binding sites for C/EBP-ß, and potential binding sites for SP-1 and AP-2 (Rhoades et al. 1992; Wedel et al. 1996; and Pilz RB, unpublished results).

Sequences mediating transcriptional repression by cGMP were identified >1 kb upstream of the transcription start sites of the GnRH and ANF receptor genes (Belsham et al. 1996; Cao et al. 1995). Transcriptional repression of the GnRH promoter was mediated by a 300 bp tissue-specific enhancer element which was sufficient to confer cGMP-mediated repression to a heterologous promoter, but the transcription factors involved were not identified (Belsham et al. 1996). Clearly, more work is necessary to define cGMP-responsive promoter elements.

4.4 Transcription Factors Targeted by cGMP-PK I

Although cGMP-PK I could theoretically regulate gene expression indirectly, via activation/inhibition of other signal transduction pathways and protein kinase/phosphatase cascades, recent evidence demonstrates that the kinase has to translocate to the nucleus to regulate transcription of the *fos* promoter, suggesting that cGMP-PK I may directly phosphorylate nuclear transcription factors [(Gudi et al. 1997) and see below]. Candidate transcription factors targeted by cGMP-PK I are those of the CRE-binding protein (CREB)/ATF and C/EBP families because (i) binding sites for these transcription factors are found in the regulatory elements of several cGMP-responsive genes (de Groot et al. 1991; Rhoades et al. 1992; Perez-Albuerne et al. 1993); (ii) sequence elements known to bind these proteins confer cGMP-responsiveness to the truncated, enhancerless (–53) *fos* promoter (Peunova and Enikolopov, 1993; Gudi et al. 1996); and (iii) increased phosphorylation of CREB/ATF-related proteins is observed in response to cGMP-PK activation (Pilz RB, unpublished results). Indeed, the transactivation potential of the chimeric transcription factor Gal4-CREB was stimulated by the combination of cGMP- and Ca^{2+}-elevating agents in PC12 cells (Peunova and Enikolopov 1993), and in cGMP-PK-transfected BHK cells, treatment with cGMP analogs alone increased the transactivation potential of Gal4-CREB (Gudi et al. 1996). In both cell types, S^{133}A mutation of CREB rendered the protein unresponsive to cGMP (Peunova and Enikolopov 1993; Gudi et al. 1996).

In several different cell types including PC12 cells, cGMP induces not only the expression of *c-fos* and *junB* mRNA, but also increases AP-1 DNA binding and transcriptional activity of AP-1-dependent reporter genes (Haby et al. 1994; Pilz et al. 1995; Thiriet et al. 1997). Two other transcription factors important for cell growth which are influenced by intracellular cGMP levels include *egr-1* and E2F-1 (Morita et al. 1997; Thiriet et al. 1997). Whether the cGMP-induced expression of *egr-1* mRNA results in increased DNA binding and/or transcriptional activity of this transcription factor remains to be determined (Thiriet et al. 1997). In hypoxic vascular SMCs, cGMP-elevating agents and cGMP analogs were found to decrease mRNA and protein expression of the transcription factor E2F-1, effects which may or may not be mediated by cGMP-PK (Morita et al. 1997).

4.5 Regulation of the c-*fos* Promoter by cGMP-PK I Requires Nuclear Translocation of the Kinase

Regulation of gene expression by cAMP-PK requires dissociation of the cAMP-PK tetramer, translocation of the free catalytic (C) subunits of cAMP-PK into the nucleus and direct phosphorylation of transcription factors such as CREB (Hagiwara et al. 1993). Although nuclear entry of the 38 kDa C-subunit of cAMP-PK occurs by passive diffusion, this appears to be the rate-limiting step in coupling hormonal stimulation of cAMP synthesis to changes in transcription of cAMP-responsive genes (Hagiwara et al. 1993). cGMP-PK with a mass of approximately 150 kDa would not be expected to enter the nucleus unless it contains a signal sequence for active transport mediated by the nuclear pore complex (Gorlich and Mattaj 1996). Indirect immunofluorescence studies of the subcellular localization of cGMP-PK in neutrophils and monocytes suggested nuclear staining by cGMP-PK-specific antibodies (Wyatt et al. 1991; Pryzwansky et al. 1995a). Recently cGMP-PK I translocation to the nucleus was observed in response to cGMP activation, and a nuclear localization signal (NLS) was identified (amino acids 404-411 of cGMP-PK Iß) which is both necessary for nuclear transport of cGMP-PK as well as sufficient for directing a heterologous protein to the nucleus (Fig. 2) (Gudi et al. 1997). Mutation of a single amino acid ($K^{407}E$) within the cGMP-PK NLS produced an enzyme which retained normal cGMP-dependent activity *in vitro*, but neither translocated to the nucleus nor transactivated the *fos* promoter in the presence of cGMP *in vivo*. In contrast, N-terminally truncated versions of cGMP-PK I with constitutive, cGMP-independent activity *in vitro*, localized to the nucleus and transactivated the *fos* promoter in the absence of cGMP (Gudi et al. 1997). The NLS appears to be cryptic in the inactive kinase and exposed through a conformational change induced by cGMP binding or by removal of the N-terminal autoin-

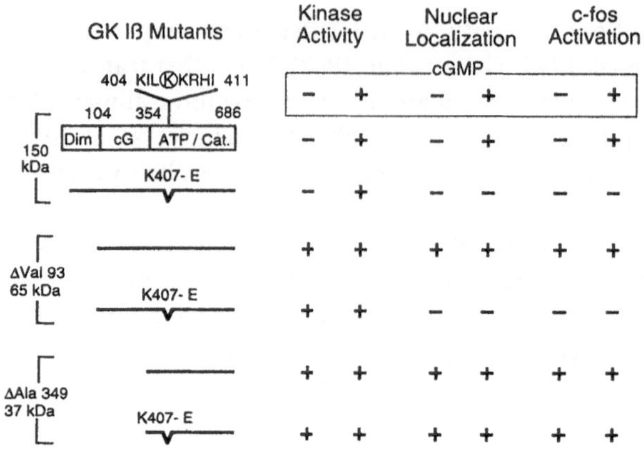

Fig. 2. Mutation (e.g. $K^{407}E$) of a putative nuclear localization signal (NLS, amino acids 404–411) in cGMP-PK Iβ inhibits its cGMP-stimulated nuclear localization and c-fos transcriptional activation. cGMP activation of cGMP-PK I appears to induce a conformational change that not only unmasks the cGMP-PK I ATP/catalytic site, but also the nearby NLS. Removal of the cGMP-PK I N-terminus (ΔVal93) containing the inhibitory region creates not only a constitutively active enzyme (not requiring cGMP for activation), but also one for which cGMP is unnecessary for nuclear localization and gene transcription. Further shortening of cGMP-PK Iβ (ΔAla349) still produces an active enzyme but it does not require the NLS for its nuclear function since it is most likely small enough to enter the nucleus by diffusion. Abbreviations used in the cGMP-PK Iβ structural domain diagram (shown at the upper left of figure): Dim dimerization domain; cG two cGMP-binding domains, ATP/Cat ATP binding and catalytic domains. Modified from Gudi et al. (1997)

hibitory domain (Chu et al. 1997; Gudi et al. 1997). These results indicate that nuclear translocation of cGMP-PK I is required for transcriptional regulation of the *fos* promoter and that cGMP-PK I most likely regulates the activity of certain transcription factors directly in the nucleus. In addition, cGMP-PK may also regulate gene expression through indirect mechanisms, e.g., by regulating cytosolic protein kinases such as c-Raf kinase (Suhasini M. et al. 1998).

4.6 Physiological Functions of cGMP-PK Likely to Involve Changes in Gene Expression

As described above, cGMP-PK has been shown to be involved in the regulation of vascular SMC differentiation. Expression of cGMP-PK I transformed de-differentiated, cGMP-PK-deficient vascular SMC to a more differentiated, contractile phenotype. In kinase-containing (but not in kinase-deficient) vascular SMCs, cGMP treatment resulted in increased expression of contractile proteins and down-regulation of extracellular matrix proteins, a pattern characteristic of the contractile phenotype (Boerth et al. 1997; Dey et al. 1998). Other important effects of cGMP on the regulation of cell growth, differentiation and apoptosis may be mediated by cGMP-PK, although in many cases this has not yet been established according to all of the criteria listed above. cGMP-PK has been implicated in the differentiation of hematopoietic and osteoblastic cells, and cGMP-induced changes in cell morphology and expression of differentiation-associated genes were seen in these cell types (Pilz et al. 1994; Inoue et al. 1995 and 1996). An important function of cGMP-PK in bone formation was demonstrated in cGMP-PK II-deficient transgenic mice which had impaired endochondral ossification and were dwarf (Pfeifer et al. 1996). Interestingly, overactive endochondral ossification and marked skeletal overgrowth was observed in transgenic mice overexpressing brain natriuretic peptide (Suda et al. 1998).

The regulation of cell growth by cGMP-PK remains controversial (Boerth et al. 1997; Yu et al. 1997; Dhaunsi and Hassid, 1996; Sugimoto et al. 1996; Cordelier et al. 1997). We have recently observed cGMP-PK-mediated inhibition of the Ras/MAP-kinase pathway which resulted in inhibition of growth factor-induced nuclear translocation of MAP-kinase and may explain the antiproliferative effect of cGMP in some cell types (Yu et al. 1997; Suhasini M et al. 1998). The role of cGMP-PK in mediating the induction or inhibition of apoptosis by cGMP in different cell types is far from clear at this point, but may also involve the regulation of gene expression (Pollman et al. 1996; Loweth et al. 1997; Kim et al. 1997). Finally, some of the functions proposed for cGMP-PK in the central nervous system imply regulation of gene expression by cGMP-PK (Peunova and Enikolopov 1993; Lincoln 1995b; Belsham et al. 1996).

5
Concluding Remarks

Diverse and complimentary experimental approaches in cGMP-PK-expressing and cGMP-PK-deficient cells and transgenic animals have established that cGMP-PK type I mediates the effects of cGMP on inhibition of platelet aggregation, inhibition of vascular SMC migration and contraction, and regulation of gene expression. Stimulation of cGMP-PK type II activates intestinal chloride transport, inhibits cAMP-stimulated renin release from juxtaglomerular cells, and plays a role in enchondral ossification. The NO/cGMP/cGMP-PK signaling cascade is present in most ECs, vascular SMCs, platelets and leukocytes, and a role of this pathway in the regulation of cell adhesion, migration and proliferation is emerging. However, many of the downstream effects of NO and cGMP are still poorly defined and the exact molecular mechanisms of NO/cGMP action, especially the cGMP-PK substrates involved, remain to be identified. A better analysis of the downstream effects of NO/cGMP/cGMP-PK will be important for our understanding of complex physiological and pathological processes such as cell growth, motility and differentiation, angiogenesis and pathological destruction of vessel walls.

Acknowledgements. M.E., S.M.L., and U.W. were supported by the Deutsche Forschungsgemeinschaft; R.B. was supported by the National Science Foundation.

References

Angel P, Karin M (1991) The role of Jun, Fos and the AP-1 complex in cell proliferation and transformation. Biochim Biophys Acta 1072:129–157

Bassenge E (1994) Coronary vasomotor responses: role of endothelium and nitrovasodilators. Cardiovasc Drugs Ther 8:601–610

Belsham DD, Wetsel WC, Mellon PL (1996) NMDA and nitric oxide act through the cGMP signal transduction pathway to repress hypothalamic gonadotropin-releasing hormone gene expression. EMBO J 15:538–547

Bischoff J (1997) Cell adhesion and angiogenesis. J Clin Invest 99:373–376

Bloch W, Forsberg E, Lentini S, Brakebusch C, Martin K, Krell HW, Weidle UH, Addicks K, Fassler R (1997) Beta 1 integrin is essential for teratoma growth and angiogenesis. J Cell Biol 139:265–278

Boerth NJ, Dey NB, Cornwell TL, Lincoln TM (1997) Cyclic GMP-dependent protein kinase regulates vascular smooth muscle cell phenotype. J Vasc Res 34:245–259

Bowler C, Neuhaus G, Yamagata H, Chua NH (1994) Cyclic GMP and calcium mediate phytochrome phototransduction. Cell 77:73–81

Brooks PC, Clark RA, Cheresh DA (1994) Requirement of vascular integrin alpha v beta 3 for angiogenesis. Science 264:569–571

Burridge K, Chrzanowska-Wodnicka M, Zhong C (1997) Focal adhesion assembly. Trends Cell Biol 7:342–347

Butt E, Geiger J, Jarchau T, Lohmann SM, Walter U (1993) The cGMP-dependent protein kinase – gene, protein, and function. Neurochem Res 18:27–42

Butt E, Eigenthaler M, Genieser HG (1994a) Rp-8-pCPT-cGMPS, a novel cGMP-dependent protein kinase inhibitor. Eur J Pharmacol 269:265–268

Butt E, Abel K, Krieger M, Palm D, Hoppe V, Hoppe J, Walter U (1994b) cAMP- and cGMP-dependent protein kinase phosphorylation sites of the focal adhesion protein VASP *in vitro* and in intact human platelets. J Biol Chem 269:14509–14517

Cao L, Wu J, Gardner DG (1995) Atrial natriuretic peptide suppresses the transcription of its guanylyl cyclase-linked receptor. J Biol Chem 270:24891–24897

Chen L, Daum G, Forough R, Clowes M, Walter U, Clowes AW (1998) Overexpression of human endothelial nitric oxide synthase in rat vascular smooth muscle cells and in balloon-injured carotid artery. Circ Res 82:862–870

Chin K, Kurashima Y, Ogura T, Tajiri H, Yoshida S, Esumi H (1997) Induction of vascular endothelial growth factor by nitric oxide in human glioblastoma and hepatocellular carcinoma cells. Oncogene 15:437–442

Chu DM, Corbin JD, Grimes KA, Francis SH (1997) Activation by cyclic GMP binding causes an apparent conformational change in cGMP-dependent protein kinase. J Biol Chem 272:31922–31928

Corbin JD, Lincoln TM (1978) Comparison of cAMP- and cGMP-dependent protein kinases. Adv Cyclic Nucleotide Res 9:159–170

Cordelier P, Esteve JP, Bousquet C, Delesque N, O'Carroll AM, Schally AV, Vaysse N, Susini C, Buscail L (1997) Characterization of the antiproliferative signal mediated by the somatostatin receptor subtype sst5. Proc Natl Acad Sci USA 94:9343–9348

Davenpeck KL, Gauthier TW, Lefer AM (1994) Inhibition of endothelial-derived nitric oxide promotes P-selectin expression and actions in the rat microcirculation. Gastroenterology 107:1050–1058

De Caterina R, Libby P, Peng HB, Thannickal VJ, Rajavashisth TB, Gimbrone MAJ, Shin WS, Liao JK (1995) Nitric oxide decreases cytokine-induced endothelial activation. Nitric oxide selectively reduces endothelial expression of adhesion molecules and proinflammatory cytokines. J Clin Invest 96:60–68

de Groot RP, Auwerx J, Karperien M, Staels B, Kruijer W (1991) Activation of junB by PKC and PKA signal transduction through a novel cis-acting element. Nucleic Acids Res 19:775–781

DeJonge HR (1981) Cyclic GMP-dependent protein kinase in intestinal brushborders. Adv Cyclic Nucleotide Res 14:315–333

Dey NB, Boerth NJ, Murphy-Ullrich JE, Chang PL, Prince CW, Lincoln TM (1998) Cyclic GMP-dependent protein kinase inhibits osteopontin and thrombospondin production in rat aortic smooth muscle cells. Circ Res 82:139–146

Dhaunsi GS, Hassid A (1996) Atrial and C-type natriuretic peptides amplify growth factor activity in primary aortic smooth muscle cells. Cardiovasc Res 31:37–47

Draijer R, Vaandrager AB, Nolte C, De Jonge HR, Walter U, van Hinsbergh VW (1995) Expression of cGMP-dependent protein kinase I and phosphorylation of its substrate, vasodilator-stimulated phosphoprotein, in human endothelial cells of different origin. Circ Res 77:897–905

Dubey RK, Jackson EK, Luscher TF (1995) Nitric oxide inhibits angiotensin II-induced migration of rat aortic smooth muscle cell. Role of cyclic-nucleotides and angiotensin 1 receptors. J Clin Invest 96:141–149

Ecker T, Gobel C, Hullin R, Rettig R, Seitz G, Hofmann F (1989) Decreased cardiac concentration of cGMP kinase in hypertensive animals. An index for cardiac vascularization? Circ Res 65:1361–1369

Eigenthaler M, Nolte C, Halbrugge M, Walter U (1992) Concentration and regulation of cyclic nucleotides, cyclic-nucleotide-dependent protein kinases and one of their major substrates in human platelets. Estimating the rate of cAMP-regulated and cGMP-regulated protein phosphorylation in intact cells. Eur J Biochem 205:471–481

Eigenthaler M, Ullrich H, Geiger J, Horstrup K, Hönig-Liedl P, Wiebecke D, Walter U (1993) Defective nitrovasodilator-stimulated protein phosphorylation and calcium regulation in cGMP-dependent protein kinase deficient human platelets of chronic myelocytic leukemia. J Biol Chem 268:13526–13531

Eigenthaler M, Shattil SJ (1996) Integrin signaling and the platelet cytoskeleton. Current Topics in Membranes 43:265–291

Elferink JGR and de Koster BM (1993) The effect of cyclic GMP and cyclic AMP on migration by electroporated human neutrophils. Eur J Pharmacol 246:157–161

Elferink JGR, VanUffelen BE (1996) The role of cyclic nucleotides in neutrophil migration. Gen Pharmacol 27:387–393

Filippov G, Bloch DB, Bloch KD (1997) Nitric oxide decreases stability of mRNAs encoding soluble guanylate cyclase subunits in rat pulmonary artery smooth muscle cells. J Clin Invest 100:942–948

Fisch TM, Prywes R, Simon MC, Roeder RG (1989) Multiple sequence elements in the c-fos promoter mediate induction by cAMP. Genes Dev 3:198–211

Francis SH, Corbin JD (1994) Structure and function of cyclic nucleotide-dependent protein kinases. Annu Rev Physiol 56:237–272

French PJ, Bijman J, Edixhoven M, Vaandrager AB, Scholte BJ, Lohmann SM, Nairn AC, DeJonge HR (1995) Isotype-specific activation of cystic fibrosis transmembrane conductance regulator-chloride channel by cGMP-dependent protein kinase II. J Biol Chem 270:26626–26631

Friedlander M, Brooks PC, Shaffer RW, Kincaid CM, Varner JA, Cheresh DA (1995) Definition of two angiogenic pathways by distinct alpha v integrins. Science 270:1500–1502

Gambaryan S, Häusler C, Markert T, Pöhler D, Jarchau T, Walter U, Haase W, Kurtz A, Lohmann SM (1996) Expression of type II cGMP-dependent protein kinase in rat kidney is regulated by dehydration and correlated with renin gene expression. J Clin Invest 98:662–670

Gambaryan S, Wagner C, Smolenski A, Walter U, Poller W, Haase W, Kurtz A, Lohmann SM (1998) Endogenous or overexpressed cGMP-dependent protein kinase inhibit cAMP-dependent renin release from rat isolated perfused kidney, microdissected glomeruli, and isolated juxtaglomerular cells. Proc Natl Acad Sci USA 95:9003–9008

Garg UC, Hassid A (1989) Nitric oxide-generating vasodilators and 8-Bromo-cyclic guanosine monophosphate inhibit mitogenesis and proliferation of cultured rat vascular smooth muscle cells. J Clin Invest 83:1774–1777

Geiger J, Nolte C, Butt E, Sage SO, Walter U (1992) Role of cGMP and cGMP-dependent protein kinase in nitrovasodilator inhibition of agonist-evoked calcium elevation in human platelets. Proc Natl Acad Sci USA 89:1031–1035

Geiger J, Nolte C, Walter U (1994) Regulation of calcium mobilization and calcium entry in human platelets by cyclic nucleotide-elevating and endothelium-derived factors. Am J Physiol 267:C236–C244

Gill GN, Holdy KE, Walton GM, Kanstein CB (1976) Purification and characterization of 3',5'-cyclic GMP-dependent protein kinase. Proc Natl Acad Sci USA 73:3918–3922

Glass DB, Cheng HC, Kemp BE, Walsh DA (1986) Differential and common recognition of the catalytic sites of the cGMP- dependent and cAMP-dependent protein kinases by inhibitory peptides derived from the heat-stable inhibitor protein. J Biol Chem 261:12166–12171

Gong JH, Renz H, Sprenger H, Nain M, Gemsa D (1990) Enhancement of tumor necrosis factor-alpha gene expression by low doses of prostaglandin E2 and cyclic GMP. Immunobiology 182:44–55

Gorlich D, Mattaj IW (1996) Nucleocytoplasmic transport. Science 271:1513–1518

Gudi T, Huvar I, Meinecke M, Lohmann SM, Boss GR, Pilz RB (1996) Regulation of gene expression by cGMP-dependent protein kinase. Transactivation of the c-fos promoter. J Biol Chem 271:4597–4600

Gudi T, Lohmann SM, Pilz RB (1997) Regulation of gene expression by cyclic GMP-dependent protein kinase requires nuclear translocation of the kinase:identification of a nuclear localization signal. Mol Cell Biol 17:5244–5254

Haby C, Lisovoski F, Aunis D, Zwiller J (1994) Stimulation of the cyclic GMP pathway by NO induces expression of the immediate early genes c-fos and junB in PC12 cells. J Neurochem 62:496–501

Hagiwara M, Brindle P, Harootunian A, Armstrong R, Rivier J, Vale W, Tsien R, Montminy MR (1993) Coupling of hormonal stimulation and transcription via the cyclic AMP- responsive factor CREB is rate limited by nuclear entry of protein kinase A. Mol Cell Biol 13:4852–4859

Hasegawa T, Takagi S, Nishimaki K, Morita K, Nakajima S (1992) Impairment of L-arginine metabolism in spontaneously hypertensive rats. Biochem Int 26:653–658

Hofmann F, Dostmann W, Keilbach A, Landgraf W, Ruth P (1992) Structure and physiological role of cGMP-dependent protein kinase. Biochim Biophys Acta 1135:51–60

Horstrup K, Jablonka B, Honig-Liedl P, Just M, Kochsiek K, Walter U (1994) Phosphorylation of focal adhesion vasodilator-stimulated phosphoprotein at Ser157 in intact human platelets correlates with fibrinogen receptor inhibition. Eur J Biochem 225:21–27

Ignarro L, Murad F (1995) Nitric oxide. Biochemistry, molecular biology and therapeutic implications. Academic Press, San Diego

Ikeda M, Kohno M, Takeda T (1995) Inhibition by cardiac natriuretic peptides of rat vascular endothelial cell migration. Hypertension 26:401–405

Inoue A, Hiruma Y, Hirose S, Yamaguchi A, Hagiwara H (1995) Reciprocal regulation by cyclic nucleotides of the differentiation of rat osteoblast-like cells and mineralization of nodules. Biochem Biophys Res Commun 215:1104–1110

Inoue A, Hiruma Y, Hirose S, Yamaguchi A, Furuya M, Tanaka S, Hagiwara H (1996) Stimulation by C-type natriuretic peptide of the differentiation of clonal osteoblastic MC3T3-E1 cells. Biochem Biophys Res Commun 221:703–707

Itoh H, Pratt RE, Dzau VJ (1990) Atrial natriuretic polypeptide inhibits hypertrophy of vascular smooth muscle cells. J Clin Invest 86:1690–1697

Itoh H, Pratt RE, Ohno M, Dzau VJ (1992) Atrial natriuretic polypeptide as a novel antigrowth factor of endothelial cells. Hypertension 19:758–761

Jarchau T, Hausler C, Markert T, Pohler D, Vanderkerckhove J, De Jonge HR, Lohmann SM, Walter U (1994) Cloning, expression, and in situ localization of rat intestinal cGMP-dependent protein kinase II. Proc Natl Acad Sci USA 91:9426–9430

Johnston HM, Morris BJ (1994) NMDA and nitric oxide increase microtubule-associated protein 2 gene expression in hippocampal granule cells. J Neurochem 63:379–382

Kalderon D, Rubin G M (1989) cGMP-dependent protein kinase genes in Drosophila. J Biol Chem 264:10738–10748

Kaupp UB (1995) Family of cyclic nucleotide gated ion channels. Curr Opin Neurobiol 5:434–442

Kawada N, Kuroki T, Uoya M, Inoue M, Kobayashi K (1996) Smooth muscle alpha-actin expression in rat hepatic stellate cell is regulated by nitric oxide and cGMP production. Biochem Biophys Res Commun 229:238–242

Kemp BE, Cheng H-C, Walsh DA (1988) Peptide inhibitors of cAMP-dependent protein kinases. Methods Enzymol 159:173–183

Kim YM, Talanian RV, Billiar TR (1997) Nitric oxide inhibits apoptosis by preventing increases in caspase-3- like activity via two distinct mechanisms. J Biol Chem 272:31138–31148

Koesling D, Bohme E, Schultz G (1991) Guanylyl cyclases, a growing family of signal-transducing enzymes. FASEB J 5:2785–2791

Kohno M, Yokokawa K, Yasunari K, Kano H, Minami M, Ueda M, Yoshikawa J (1997) Effect of natriuretic peptide family on the oxidized LDL-induced migration of human coronary artery smooth muscle cells. Circ Res 81:585–590

Kuo JF, Greengard P (1970) Cyclic nucleotide-dependent protein kinase VI. Isolation and partial purification of a protein kinase activated by guanosine 3',5'-monophosphate. J Biol Chem 245:2493–2498

Kuo JF, Davis CW, Tse J (1976) Depressed cardiac cyclic GMP-dependent protein kinase in spontaneously hypertensive rats and its further depression by guanethidine. Nature 261:335–336

Kurose I, Kubes P, Wolf R, Anderson DC, Paulson J, Miyasaka M, Granger DN (1993) Inhibition of nitric oxide production. Mechanisms of vascular albumin leakage. Circ Res 73:164–171

Lau YT, Ma WC (1996) Nitric oxide inhibits migration of cultured endothelial cells. Biochem Biophys Res Commun 221:670–674

Lee MR, Li L, Kitazawa T (1997) Cyclic GMP causes Ca^{2+} desensitization in vascular smooth muscle by activating the myosin light chain phosphatase. J Biol Chem 272:5063–5068

Lincoln TM, Dills WL, Corbin JD (1977) Purification and subunit composition of guanosine 3':5'-monophosphate-dependent protein kinase from bovine lung. J Biol Chem 252:4269–4275

Lincoln TM, Komalavilas P, Boerth NJ, MacMillan-Crow LA, Cornwell TL (1995a) cGMP signaling through cAMP- and cGMP-dependent protein kinases. Adv Pharmacol 34:305–322

Lincoln TM (1995b) Nitric oxide in the nervous system. Academic Press, San Diego 51 pp

Lohmann SM, Vaandrager AB, Smolenski A, Walter U, DeJonge HR (1997) Distinct and specific functions of cGMP-dependent protein kinases. TIBS 22:307–312

Loweth AC, Williams GT, Scarpello JH, Morgan NG (1997) Evidence for the involvement of cGMP and protein kinase G in nitric oxide-induced apoptosis in the pancreatic B-cell line, HIT-T15. FEBS Lett 400:285–288

Markert T, Krenn V, Leebmann J, Walter U (1996) High expression of the focal adhesion- and microfilament-associated protein VASP in vascular smooth muscle and endothelial cells of the intact human vessel wall. Basic Res Cardiol 91:337–343

Matsuoka H, Nakata M, Kohno K, Koga Y, Nomura G, Toshima H, Imaizumi T (1996) Chronic L-arginine administration attenuates cardiac hypertrophy in spontaneously hypertensive rats. Hypertension 27:14–18

McEver RP, Cummings RD (1997) Perspectives series:cell adhesion in vascular biology. Role of PSGL-1 binding to selectins in leukocyte recruitment. J Clin Invest 100:485–491

Mecham RP, Levy BD, Morris SL, Madaras JG, Wrenn DS (1985) Increased cyclic GMP levels lead to a stimulation of elastin production in ligament fibroblasts that is reversed by cyclic AMP. J Biol Chem 260:3255–3258

Mönks D, Lange V, Silber RE, Markert T, Walter U, Nehls V (1998) Expression of cGMP-dependent protein kinase I and its substrate VASP in neointimal cells of the injured rat carotid artery. Eur J Clin Invest, 28:416–423

Morita T, Mitsialis SA, Koike H, Liu Y, Kourembanas S (1997) Carbon monoxide controls the proliferation of hypoxic vascular smooth muscle cells. J Biol Chem 272:32804–32809

Murphy-Ullrich JE, Pallero MA, Boerth N, Greenwood JA, Lincoln TM, Cornwell TL (1996) Cyclic GMP-dependent protein kinase is required for thrombospondin and tenascin mediated focal adhesion disassembly. J Cell Sci 109:2499–2508

Niu XF, Smith CW, Kubes P (1994) Intracellular oxidative stress induced by nitric oxide synthesis inhibition increases endothelial cell adhesion to neutrophils. Circ Res 74:1133–1140

Niu XF, Ibbotson G, Kubes P (1996) A balance between nitric oxide and oxidants regulates mast cell- dependent neutrophil-endothelial cell interactions. Circ Res 79:992–999

Nolte C, Eigenthaler M, Schanzenbächer P, Walter U (1991) Endothelial cell-dependent phosphorylation of a platelet protein mediated by cAMP- and cGMP-elevating factors. J Biol Chem 266:14808–14812

Orstavik S, Natarajan V, Tasken K, Jahnsen T, Sandberg M (1997) Characterization of the human gene coding for the type I alpha and type I beta cGMP-dependent protein kinase. Genomics 42:311–318

Osborne KA, Robichon A, Burgess E, Butland S, Shaw RA, Coulthard A, Pereira HS, Greenspan RJ, Sokolowski MB (1996) Natural behavior polymorphism due to a cGMP-dependent protein kinase of Drosophila. Science 277:834–836

Papapetropoulos A, Desai KM, Rudic RD, Mayer B, Zhang R, Ruiz-Torres MP, Garcia-Cardena G, Madri JA, Sessa WC (1997a) Nitric oxide synthase inhibitors attenuate transforming-growth-factor- beta 1-stimulated capillary organization in vitro. Am J Pathol 150:1835–1844

Papapetropoulos A, Garcia-Cardena G, Madri JA, Sessa WC (1997b) Nitric oxide production contributes to the angiogenic properties of vascular endothelial growth factor in human endothelial cells. J Clin Invest 100:3131–3139

Parenti A, Morbidelli L, Cui XL, Douglas JG, Hood JD, Granger HJ, Ledda F, Ziche M (1998) Nitric oxide is an upstream signal of vascular endothelial growth factor-induced extracellular signal-regulated kinase 1/2 activation in postcapillary endothelium. J Biol Chem 273:4220–4226

Peng HB, Libby P, Liao JK (1995) Induction and stabilization of I kappa B alpha by nitric oxide mediates inhibition of NF-kappa B. J Biol Chem 270:14214–14219

Perez-Albuerne ED, Schatteman G, Sanders LK, Nathans D (1993) Transcriptional regulatory elements downstream of the JunB gene. Proc Natl Acad Sci USA 90:11960–11964

Peunova N, Enikolopov G (1993) Amplification of calcium-induced gene transcription by nitric oxide in neuronal cells. Nature 364:450–453

Pfeifer A, Nurnberg B, Kamm S, Uhde M, Schultz G, Ruth P, Hofmann F (1995) Cyclic GMP-dependent protein kinase blocks pertussis-toxin.sensitive hormone receptor signaling pathways in Chinese hamster ovary cells. J Biol Chem 270:9052–9059

Pfeifer A, Aszodi A, Seidler U, Ruth P, Hofmann F, Fässler R (1996) Intestinal secretory defects and dwarfism in mice lacking cGMP-dependent protein kinase II. Science 274:2082–2086

Pfeifer A, Klatt P, Massberg S, Ny L, Sausbier M, Hirneiss C, Wang G-X, Korth M, Aszodi A, Andersson K-E, Krombach F, Mayerhofer A, Ruth P, Fässler R, Hofmann F (1998) Defective smooth muscle regulation in cGMP kinase I-deficient mice. EMBO J 17:3045–3051

Pilz RB, Berjis M, Idriss SD, Scheele JS, Suhasini M, Gao L, Scheffler IE, Boss GR (1994) Isolation and characterization of HL-60 cells resistant to nitroprusside-induced differentiation. J Biol Chem 269:32155–32161

Pilz RB, Suhasini M, Idriss S, Meinkoth JL, Boss GR (1995) Nitric oxide and cGMP analogs activate transcription from AP-1- responsive promoters in mammalian cells. FASEB J 9:552–558

Pollman MJ, Yamada T, Horiuchi M, Gibbons GH (1996) Vasoactive substances regulate vascular smooth muscle cell apoptosis. Countervailing influences of nitric oxide and angiotensin II. Circ Res 79:748–756

Pryzwansky KB, Wyatt TA, Nichols H, Lincoln TM (1990) Compartmentalization of cyclic GMP-dependent protein kinase in formyl- peptide stimulated neutrophils. Blood 76:612–618

Pryzwansky KB, Kidao S, Wyatt TA, Reed W, Lincoln TM (1995a) Localization of cyclic GMP-dependent protein kinase in human mononuclear phagocytes. J Leukoc Biol 57:670–678

Pryzwansky KB, Wyatt TA, Lincoln TM (1995b) Cyclic guanosine monophosphate-dependent protein kinase is targeted to intermediate filaments and phosphorylates vimentin in A23187-stimulated human neutrophils. Blood 85:222–230

Radomski MW, Palmer RM, Moncada S (1987) The role of nitric oxide and cGMP in platelet adhesion to vascular endothelium. Biochem Biophys Res Commun 148:1482–1489

Reinhard M, Jarchau T, Walter U (1998) VASP. In:Guidebook to the cytoskeletal and motor proteins (Eds. Kreis T, Vale R), Oxford University Press, in press

Rhoades KL, Golub SH, Economou JS (1992) The regulation of the human tumor necrosis factor alpha promoter region in macrophage, T cell, and B cell lines. J Biol Chem 267:22102–22107

Ruth P, Wang GX, Boekhoff I, May B, Pfeifer A, Penner R, Korth M, Breer H, Hofmann F (1993) Transfected cGMP-dependent protein kinase suppresses calcium transients by inhibition of inositol 1,4,5-trisphosphate production. Proc Natl Acad Sci USA 90:2623–2627

Rybalkin SD, Beavo J (1996) Multiplicity within cyclic nucleotide phosphodiesterases. Biochem Soc Trans 24:1005–1009

Sandberg M, Natarajan V, Ronander I, Kalderon D, Walter U, Lohmann SM, Jahnsen T (1989) Molecular cloning and predicted full-length amino acid sequence of the type Iß isozyme of cGMP-dependent protein kinase from human placenta. Tissue distribution and developmental changes in rat. FEBS Lett 225:321–329.

Sansom SC, Stockand JD, Hall D, Williams B (1997) Regulation of large calcium-activated potassium channels by protein phosphatase 2A. J Biol Chem 272:9902–9906

Sarkar R, Meinberg EG, Stanley JC, Gordon D, Webb RC (1996) Nitric oxide reversibly inhibits the migration of cultured vascular smooth muscle cells. Circ Res 78:225–230

Schmidt HHHW, Walter U (1994) NO at work. Cell 78, 919–925.

Schwartz SM (1997) Smooth muscle migration in atherosclerosis and restenosis. J Clin Invest 100:S87–S89

Smolenski A, Burkhardt M, Eigenthaler M, Butt E, Gambaryan S, Lohmann SM, Walter U (1998) Functional analysis of cGMP-dependent protein kinases I and II as mediators of NO/cGMP effects. Naunyn-Schmiedeberg's Arch Pharmacol 358:134–139

Soff GA, Cornwell TL, Cundiff DL, Gately S, Lincoln TM (1997) Smooth muscle expression of type I cyclic GMP-dependent protein kinase is suppressed by continuous exposure to nitrovasodilators, theophylline, cyclic GMP, and cyclic AMP. J Clin Invest 100:2580–2587

Stockert RJ, Ren Q (1997) Cytoplasmic protein mRNA interaction mediates cGMP-modulated translational control of the asialoglycoprotein receptor. J Biol Chem 272:9161–9165

Suda M, Ogawa Y, Tanaka K, Tamura N, Yasoda A, Takigawa T, Uehira M, Nishimoto H, Itoh H, Saito Y, Shiota K, Nakao K (1998) Skeletal overgrowth in transgenic mice that overexpress brain natriuretic peptide. Proc Natl Acad Sci USA 95:2337–2342

Sugimoto T, Haneda M, Togawa M, Isono M, Shikano T, Araki S, Nakagawa T, Kashiwagi A, Guan KL, Kikkawa R (1996) Atrial natriuretic peptide induces the expression of MKP-1, a mitogen-activated protein kinase phosphatase, in glomerular mesangial cells. J Biol Chem 271:544–547

Suhasini M, Li H, Lohmann SM, Boss GR, Pilz RB (1998) cGMP-dependent protein kinase inhibits the ras/mitogen-activated protein kinase pathway. Mol Cell Biol, in press

Sundqvist T, Forslund T, Bengtsson T, Axelsson KL (1994) S-nitroso-N-acetylpenicillamine reduces leukocyte adhesion to type I collagen. Inflammation 18:625–631

Takahashi M, Ikeda U, Masuyama J, Funayama H, Kano S, Shimada K (1996) Nitric oxide attenuates adhesion molecule expression in human endothelial cells. Cytokine 8:817–821

Takio K, Wade RD, Smith SB, Krebs EG, Walsh KA, Titani K (1984) Guanosine cyclic 3',5'-phosphate-dependent protein kinase, a chimeric protein homologous with two separate protein families. Biochemistry 23:4207–4218

Tetsuka T, Daphna-Iken D, Miller BW, Guan Z, Baier LD, Morrison AR (1996) Nitric oxide amplifies interleukin 1-induced cyclooxygenase-2 expression in rat mesangial cells. J Clin Invest 97:2051–2056

Thiriet N, Esteve L, Aunis D, Zwiller J (1997) Immediate early gene induction by natriuretic peptides in PC12 phaeochromocytoma and C6 glioma cells. Neuroreport 8:399–402

Tsao PS, Buitrago R, Chan JR, Cooke JP (1996) Fluid flow inhibits endothelial adhesiveness. Nitric oxide and transcriptional regulation of VCAM-1. Circulation 94:1682–1689

Uberall F, Werner-Felmayer G, Schubert C, Grunicke HH, Wachter H, Fuchs D (1994) Neopterin derivatives together with cyclic guanosine monophosphate induce c-fos gene expression. FEBS Lett 352:11–14

Uhler M D (1993) Cloning and expression of a novel cyclic GMP-dependent protein kinase from mouse brain. J Biol Chem 268:13586–13591

Vaandrager AB, DeJonge HR (1996) Signalling by cGMP-dependent protein kinases. Mol Cell Biochem 157:23–30

Vaandrager AB, Tilly BC, Smolenski A, Schneider-Rasp S, Bot AGM, Edixhoven M, Scholte BJ, Jarchau T, Walter U, Lohmann SM, Poller WC, DeJonge HR (1997) cGMP-stimulation of cystic fibrosis transmembrane conductance regulator Cl^-

channels co-expressed with cGMP-dependent protein kinase type II but not type Iß. J Biol Chem 272:4195–4200

Vaandrager AB, Smolenski A, Tilly BC, Housmuller AB, Ehlert EME, Bot AGM, Edixhoven M, Boomaars WEM, Lohmann SM, DeJonge HR (1998) Membrane targeting of cGMP-dependent protein kinase is required for cystic fibrosis transmembrane conductance regulator Cl^- channel activation. Proc Natl Acad Sci USA 95:1466–1471

Vemulapalli S, Chiu PJ, Kurowski S, Brown A, Hartman BD, Leach MW (1996) In vivo inhibition of platelet adhesion by a cGMP-mediated mechanism in balloon catheter injured rat carotid artery. Pharmacology 52:235–242

Venturini CM, Weston LK, Kaplan JE (1992) Platelet cGMP, but not cAMP, inhibits thrombin-induced platelet adhesion to pulmonary vascular endothelium. Am J Physiol 263:H606–H612

Vo NK, Gettemy JM, Coghlan VM (1998) Identification of cGMP-dependent protein kinase anchoring proteins (GKAPs). Biochem Biophys Res Commun 246:831–835

Waldmann R, Walter U (1989) Cyclic nucleotide elevating vasodilators inhibit platelet aggregation at an early step of the activation cascade. Eur J Pharmacol 159:317–320

Walter U (1984) Cyclic GMP-regulated enzymes and their possible physiological functions. Adv Cycl Nucl Prot Phosphor Res 17:249–258

Walter U (1989) Physiological role of cGMP and cGMP-dependent protein kinase in the cardiovascular system. Rev Physiol Biochem Pharmacol 113:41–88

Wang GR, Zhu Y, Halushka PV, Lincoln TM, Mendelsohn ME (1998) Mechanism of platelet inhibition by nitric oxide:*In vivo* phosphorylation of thromboxane receptor by cyclic GMP-dependent protein kinase. Proc Natl Acad Sci USA 95:4888–4893

Wang S, Yan L, Wesley RA, Danner RL (1997) Nitric oxide increases tumor necrosis factor production in differentiated U937 cells by decreasing cyclic AMP. J Biol Chem 272:5959–5965

Wedel A, Sulski G, Ziegler-Heitbrock HW (1996) CCAAT/enhancer binding protein is involved in the expression of the tumour necrosis factor gene in human monocytes. Cytokine 8:335–341

Wedel BJ, Garbers DL (1997) New insights on the functions of the guanylyl cyclase receptors. *FEBS Lett* 410:29–33

Wernet W, Flockerzi V, Hofmann F (1989) The cDNA of the two isoforms of bovine cGMP-dependent protein kinase. FEBS Lett 251:191–196

Wu CC, Ko FN, Teng CM (1997) Inhibition of platelet adhesion to collagen by cGMP-elevating agents. Biochem Biophys Res Commun 231:412–416

Wu X, Haystead TAJ, Nakamoto RK, Somlyo AV, Somlyo AP (1998) Acceleration of myosin light chain dephosphorylation and relaxation of smooth muscle by telokin. Synergism with cyclic nucleotide-activated kinase. J Biol Chem 273:11362–11369

Wyatt TA, Lincoln TM, Pryzwansky KB (1991) Vimentin is transiently co-localized with and phosphorylated by cyclic GMP-dependent protein kinase in formyl-peptide-stimulated neutrophils. J Biol Chem 266:21274–21280

Yamada KM and Geiger B (1997) Molecular interactions in cell adhesion complexes. Curr Opin Cell Biol 9:76–85

Yoshida T, Hayashi M, Monkawa T, Saruta T (1996) Regulation of obese mRNA expression by hormonal factors in primary cultures of rat adipocytes. Eur J Endocrinol 135:619–625

Yu SM, Hung LM, Lin CC (1997) cGMP-elevating agents suppress proliferation of vascular smooth muscle cells by inhibiting the activation of epidermal growth factor signaling pathway. Circulation 95:1269–1277

Zhou XB, Ruth P, Schlossmann J, Hofmann F, Korth M (1996) Protein phosphatase 2A is essential for the activation Ca^{2+}-activated K^{+} currents by cGMP-dependent protein kinase in tracheal smooth muscle and Chinese hamster ovary cells. J Biol Chem 271:19760–19767

Ziche M, Morbidelli L, Choudhuri R, Zhang HT, Donnini S, Granger HJ, Bicknell R (1997) Nitric oxide synthase lies downstream from vascular endothelial growth factor-induced but not basic fibroblast growth factor-induced angiogenesis. J Clin Invest 99:2625–2634